Giving Away the Farm

How Kindness, Critters and Yarn Knit a Community Together

Cindy Telisak

Good Shepherd Publishing

Copyright © 2014 by Cindy Telisak. All rights reserved.

Cover photography © Jennifer Jurek Photography
Interior photography © Cindy Telisak, except where noted.
"Barbed Wire Cowl" © Laurie Beardsley. Used with Permission.
Published by Good Shepherd Publishing, Parker, Texas, USA

No part of this publication may be reproduced or transmitted in any form or by any means, electronic or mechanical, including photocopying, recording or any other information storage and retrieval system, without the written permission of the publisher.

This book is available at discounts in bulk quantities for sales or promotional use. For general information on our other products or services or for technical support, please contact our Customer Care Department, at Good Shepherd Publishing, 4308 Church Lane, Parker, TX 75002. For more information about GSP products, visit our website at www.goodshepherdpublishing.com.

Editor: Elizabeth Green Musselman
Interior and front cover design: Suzanne Crolley
Back cover design: Elizabeth Green Musselman

Library of Congress Catalog Control Number 2014919609
ISBN 978-0-9907808-1-6 (paper)

Published in the United States of America

*For my husband, Ted,
and my daughter, Emma,
who put up with a lot of "roughing it"
once we left the suburbs,
and for the Barnies and critters
who taught me to be a shepherd.*

Contents

Foreword vii
Special Thanks ix

ONE: Welcome to the Farm. 14
TWO: For the Love of Sheep 20
THREE: The Paradigm: Stewardship vs. Ownership 24
FOUR: The Fiber Affair 30
FIVE: At Last, A Shepherd 38
SIX: The Siren Call of Yarn and Critters. 48
SEVEN: The Fiber CSA: A Better Way 54
EIGHT: The Must-Have Heart Conditions:
       Hospitality, Acceptance and Compassion 60
NINE: The Most Unlikely Leader. 70
TEN: Shepherd Interrupted 80
ELEVEN: Generosity and Learning to Receive 90
TWELVE: Spreading the Community beyond the LRB 106
THIRTEEN: Finding Home. 120
FOURTEEN: Community Killers 128
FIFTEEN: My Parting Promise 136

Endnotes 139
About the Author. 140

foreword

A few years ago, I found myself in the unenviable position of being hopelessly over-committed, exhausted and way behind schedule on a manuscript that was due to my publisher practically immediately. I was burned out, unmotivated and completely overwhelmed.

I decided that the only solution to my problem was to flee to my native Texas for two weeks in hopes that I could actually get some writing done at my mom's house without all the distractions of running a sheep farm and a small business.

I must have mentioned on my blog that I would be traveling to Texas, because just before I left for my trip, my assistant forwarded me an email from a woman in Dallas who wanted to set up a meeting with me while I was in town. She wanted to talk about how I set up my business and made it so successful so quickly.

My answer was an unequivocal "NOPE!" I was going to Texas to write and to dig myself out of the hole I was in. I had no time to help anyone at the moment. I needed to help myself.

And besides, I had received a half dozen emails a week like this one from sheep farmers who wanted to know how to replicate my business plan. The truth was, when it came right down to it, not one of them had actually wanted to do the work necessary to make a business like mine succeed.

My assistant sent the woman back a polite email letting her know that I wouldn't have time to meet with her and almost immediately received a reply from the woman asking if it might be possible for her to drop off lunch at my mom's house and spend 15 minutes with me. "After all," she said, "she has to eat."

Giving Away the Farm

This follow-up email charmed me a bit, but it also wore me down. "Tell her I will meet with her for half an hour at Barnes and Noble," I instructed my assistant. "No food." Pretty much immediately after hitting "send," I started regretting changing my mind.

And then an amazing thing happened. Cindy Telisak walked into the bookstore so full of enthusiasm and energy that she caught me completely off guard. I don't think I've ever met anyone who was so ready to learn, or so grateful for the opportunity.

Half an hour turned into nearly three hours as I walked Cindy through my business plan. She took copious notes and asked smart questions. But the most important thing she did was rekindle my own excitement about my business.

See, I had become so wrapped up in the minutia of running a successful business that I had completely forgotten about all the joy. I had allowed the thing that I was most proud of—starting the first Yarn CSA in the country and building it into a successful and ever-growing business—to become a burden, a drudgery, something to be borne.

Passing everything I knew on to Cindy Telisak was as rewarding for me as it was for her—maybe more so. And watching Cindy's Jacob's Reward Farm take its first shaky steps and then go on to become a thriving, sustainable and beloved business has been one of the greatest joys of my professional life.

Cindy and I are about as different as two people can be. We are pretty much on the opposite side of just about every issue you can think of, and yet there is one thing that we agree on wholeheartedly and without reservation: the best way to create something good is to give it away.

You want love, respect, success, kindness, help? Then you need to give as much of those things as you can. You need to give them when you are feeling great about yourself but you also need to give them when you are feeling burned out, overwhelmed and exhausted.

Since that fateful meeting with Cindy all those years ago, I have had many business consultants and experts tell me that I never should have helped other farmers set up Yarn CSAs. That I would never build the business I wanted if I kept helping out the competition. That I was giving away the store, taking money out of my own pocket.

I just shake my head and laugh. Because what I have given away has come back to me a hundred times over. My pockets are doing just fine and my heart? Well, it's overflowing.

The world can learn a lot from Cindy Telisak, and it has nothing to do with sheep.

—Susan Gibbs
Juniper Moon Farm
www.fiberfarm.com
Author of *Modern Country Knits: 30 Designs from Juniper Moon Farm*

Special Thanks

This little book is the product of an enormous amount of love and effort from dozens of people—even this extensive list doesn't completely cover it. I'm delighted to thank:

My family, Ted and Emma, who took the plunge with me to live a country life.

Joan and Fred Horak, who taught me many, many lessons as my shepherd mentors.

Susie Gibbs, who generously shared a strong but gentle business model and lives it out to this day.

Seth Godin, whose writing inspires and compels me to be brave and to "go be remarkable."

My church family, who kept the whole project in their prayers for months.

Jill Cheeks, my coach at Mission Publishing, who walked me down the path she has traveled many times, but where I had never been. The resources and encouragement from Mission Publishing have been invaluable.

Michelle Knoerzer, a can-do friend who always says yes when I need a hand.

Gail Stryker, my suburban farmer friend who inspires me to keep it natural and real, and who alone will help me bathe my chickens.

Laurie McIntyre, who turned on the light bulb for me that fateful day in the garden, and who is always in my corner.

My editor, advisor and back cover designer, Elizabeth Green Musselman, who kept me focused and helped me see my writing with her kind and graceful eyes.

My front cover and interior book designer, Suzanne Crolley, whose artistry I have admired for over thirty years.

My Beta readers and proofers, especially Sarah Carpenter and Mei Lin Turner, who took a good bit of time pouring over the manuscript, and who helped me overcome my Oxford comma fixation.

My generous Kickstarter backers, without whom there would be no book. The following people ponied up their support not only with their hearts, but with their wallets, and I'll be forever in their debt:

Al Friedar, Alice Curtis, Allison Clark, Allison Conkel, Alta Mantsch, Andy Perkins, Ann M Braly, Ann Mayes, Anna Branner, Anna Hulse, Ashley Ammons Wolfe, Beatriz Beltran, Betsy Abney, Bree Penninger, Bryan Hunter, Candi Summers, Cathie Mercer, Chrissy Howard, Christine O'Hara, Dawn Edwards Bahr, Deborah, Denise Romine, Denise Royal, Diane Poston, D'Wanna Whitener, Elizabeth Crecelius, Eunice Lytle, Gay Getz, Greta Poulsen, Hannah Fellows, Jane Nearing, Janice Lynn, Jeanie Collins, Jennifer Jurek, Jennifer Marshall, Jill Clark, JM Hannigan, Joan Thurber, Joanne Lumsden, John Reyes, John Suan, Joni Lawver, Joshua Garrett, Judith Bourgeois, Julie George, Karelin Seitz, Karen Dungey, Kate Culbertson, Kelli Ders, Kenneth Roe, Kris Ballestro, Kristin Ivers, Laura Stevenson, Laurie Beardsley, Laurie McIntyre, Lisa Lilly, Lisa Randolph, Lisa Woof, Lutie Larsen, Manda Sims, Mary Earle, Matthew Monroe, Mei Lin Turner, Melissa Barker, Melissa Delgado, Melissa Rice, Mickey Perdue, Misty Hartley Urech, Patricia Walters, Patrick Neil, Peggy Ann Ralston, Peggy Urban, Phebe Phillips Hargrove, Raquel Batchelor, Rita Allen, Sara Cotton, Sarah Carpenter, Sergio Arantini, SR, Sue Lowe, Suzanne Smith Collier, Suzie Sugrue, Tasha Patterson, Taya Schram, Ted Woodward-Partridge, Teresa Morris, Terri Barnett, Terri Mitchell, Terri Taylor, Tim and Cyndi Daugherty, Torre Taylor, Valerie L Wenger,

Special Thanks

Victoria Reyes, Victoria Safriet, Vinton Rafe McCabe and Colonel Walter Knoerzer.

And to all the Barnies—the regulars and the few-timers—who have made the Little Red Barn one of the most welcoming places on earth.

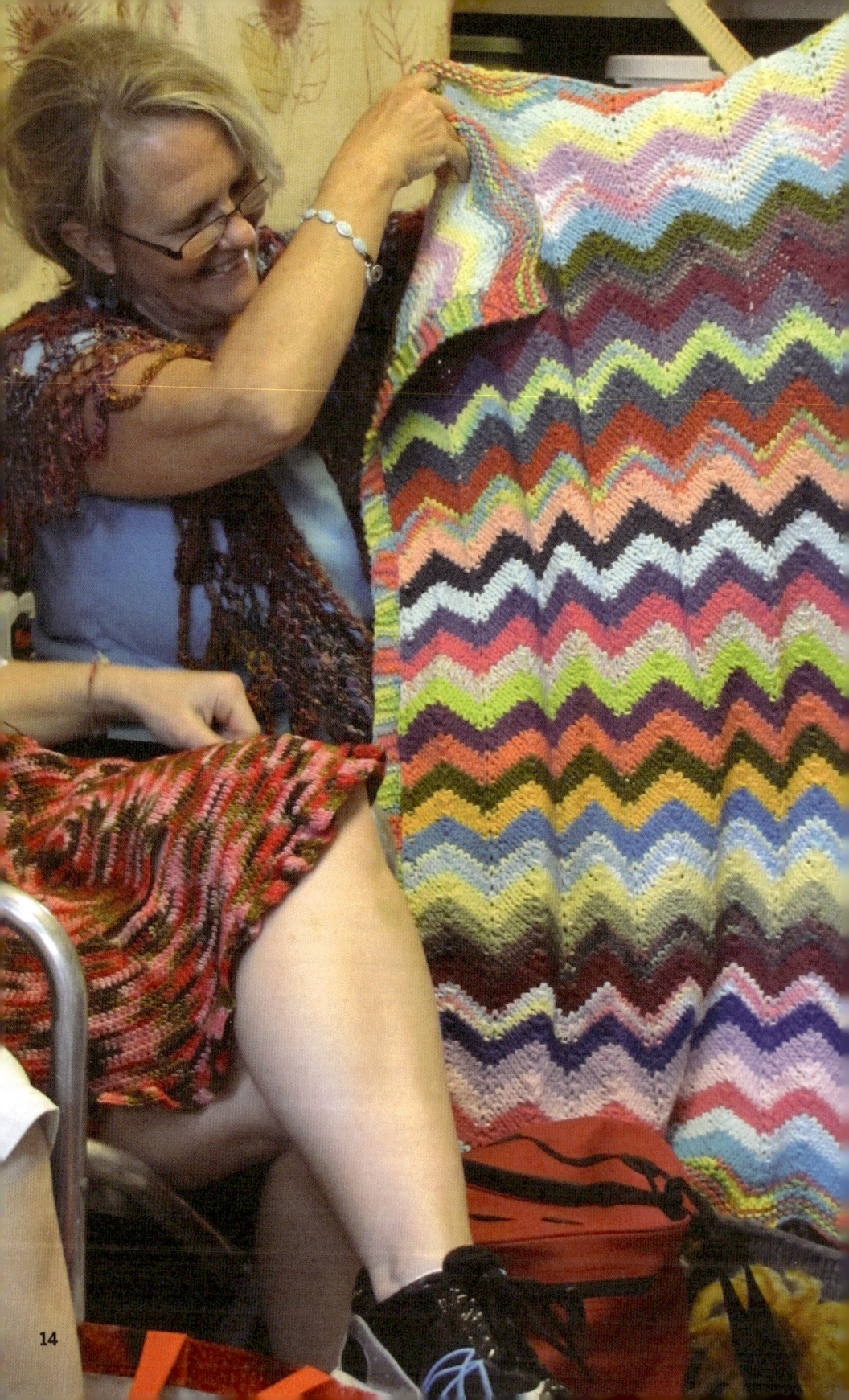

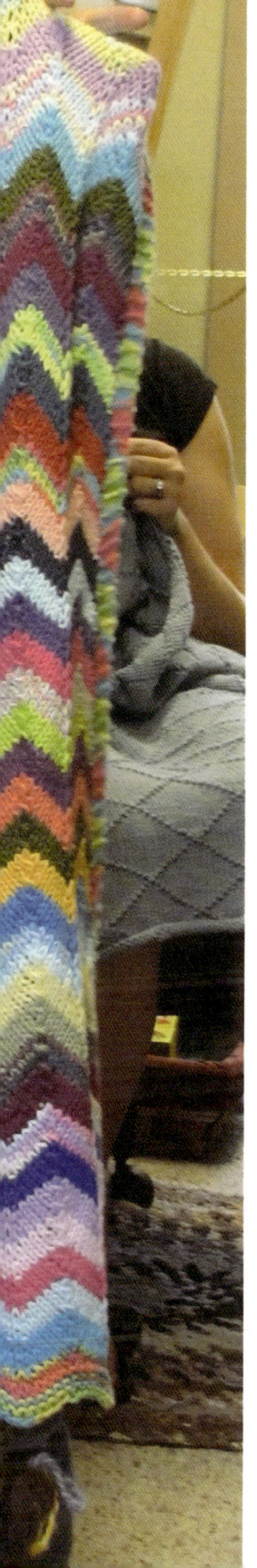

CHAPTER ONE

Welcome to the farm

I love the animals, from the alpacas that nibble at my buttons and give kisses, to the dogs that can never get enough attention. I love that my husband loves it there and likes me to go to the farm with or without him. I love the dye days where anything goes and I get to help teach people and direct traffic and beautiful stuff is made.

I love to see all the cool stuff people make.

—RITA, BARNIE

Giving Away the Farm

If you're a knitter, crocheter, spinner or other yarn crafter, and if you've ever belonged to a group of like-minded souls, I wrote this book for you.

Perhaps your group is like ours at the Little Red Barn: you feel welcome and accepted, you are encouraged in your craft, you are celebrated for your successes and victories, you are challenged to take new risks and you develop deep and lasting friendships built on trust and shared passions. You make your group meetings an anchor in your monthly schedule. You spend time online with the members of your group between meetings. You concoct off-site excuses to get together with these friends, for even more crafting time together or just to hang out. Your spouse or significant other encourages you to go because "you come back so much nicer." Your community has become a very large part of your life.

Or maybe your group is not at all like this, but you wish it were.

I'm going to share with you what to look for in a crafting community, and even help you build it if it's not already there. I'm focusing on yarn crafting because that's what I

know, and because the very nature of yarn and fiber forms the foundation for a particularly special kind of community. Seek out—and help build—the characteristics of a healthy, nurturing group, and you'll find a great treasure that will enrich your life. Collaborative creativity elevates your experience in ways that solo work cannot; iron sharpens iron. Synergy is a powerful force that can be harnessed in community when friends build on each others' innovation and energy. It's worth finding a really great group where you click. No substitute for authentic friendship exists for humans, so identifying an open, vibrant clan is really just an exercise in self-care. Artists may need support even more than the average person does because tapping into our creative juices requires emotional tightrope walking and some separation from our everyday lives. Frankly, yarn artists who have never sought out the company of others just don't know what they're missing.

It's good to know what to avoid in a crafting group, so I'll also talk about red flags to watch out for in your quest for a creative home. But just as bank tellers learn how to spot

counterfeit currency by studying *the real thing* rather than fakes, so too, learning about healthy groups will make the counterfeits stand out in stark contrast.

Jacob's Reward Farm is my home. I live here with my husband and teenage daughter, but we also host guests of all kinds nearly every week. The knitters and crocheters who visit here a lot tell me that this farm is different, that this fiber community is different from places they've visited before. Here, they feel loved and accepted regardless of their race, creed, political bent, physical appearance or artistic quirkiness. Whereas in other groups they might have felt judged or just a bit uncomfortable, here they know they belong. Here, they know they can be themselves without fear of backlash or snubbing of any kind. Some describe the farm as the "home they didn't know they were looking for." Why is Jacob's Reward special in this way? What happens here that is too often absent in other groups? How did we manage to create such a safe, welcoming place?

Frankly, I was surprised to learn from some of our community members that this particular group of friends enjoys a unique ability to get along and thrive together. To me, this is how groups ought to behave and operate. Alas, many of our friends reported shabby treatment in other groups—sometimes subtle and sometimes overt. In all cases, such treatment was regrettable and unnecessary. Because these friends matter so much to me, I began investigating what we might be doing well here so that we could continue to mature as a healthy community and not fall into any of the traps that seem to beset other groups. I believe strongly that if you don't live in ways intentionally designed to keep you headed toward your goal, you can wander down paths you never intended to travel. We have a good thing going at Jacob's Reward, and I want to be sure to protect and nurture it. I stepped back and began evaluating our "corporate identity" from the ground up, beginning with my fundamental worldview. After all, the buck stops here.

Obviously, no community is perfect—we all have room to improve and grow—but if you find a place to land that looks anything at all like the crazy bunch at the Little Red Barn (or LRB, as we have nicknamed it), you will never want to leave. Sit back, kick your shoes off and relax. You're among friends.

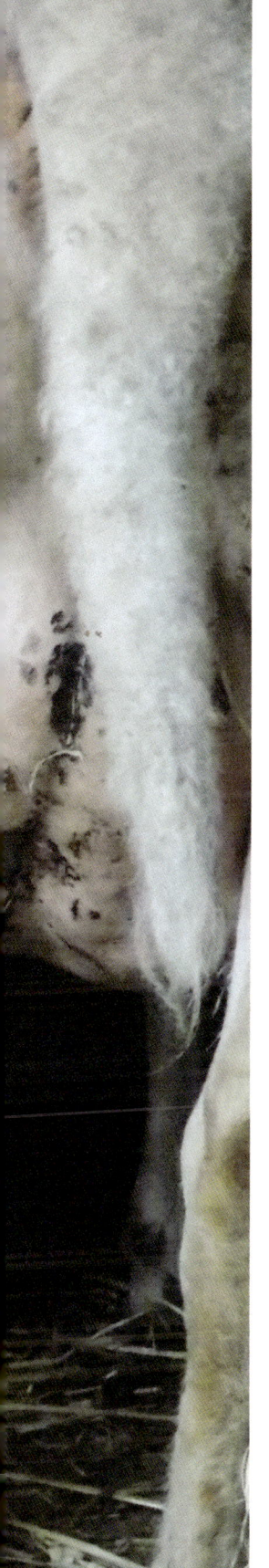

CHAPTER TWO

For the Love of Sheep

I love the farm because it brings an inner peace and tranquility that I thought I had lost. The farm brings a new purpose to my life.

—Victoria, Barnie

Giving Away the Farm

I doubt you picked up this book to get a lesson on the Gospel of John or a treatise on Jesus, the Good Shepherd. But the fact is, no telling of the story of this farm would make sense without it. Christianity, and these scriptures in particular, account for why I am who I am, and why the farm has become what it is. So bear with me for just a minute.

Every Sunday of my entire life, and many days in between, I've spent time worshiping with other Christians and reading about the life of Jesus. Jesus calls himself the Good Shepherd, and talks in detail about what makes a good shepherd and what makes a bad one. He explains how sheep behave and how we are like them—fearful, self-destructive and prone to wander into danger. He talks about how much a good shepherd loves his sheep and what he would do for them if they were threatened. He is also described as "the Lamb of God who takes away the sins of the world," the fulfillment of the Jewish practice of sacrificing sheep in the temple. Jesus and sheep are inextricably linked.

I know this is why I've always loved sheep. I've always felt loved, valued and cared for by God just like a lamb would feel in the care of a devoted shepherd. I love Jesus for volunteering to become the "sacrificial Lamb" who takes away my sins. Because of His model, there's very little about my life as a shepherd that doesn't help me understand my faith and my Lord more and more every day.

Because of this sure knowledge that I am loved and valued by my Great Shepherd, my natural response is to love others in return. It's my deepest longing and passion to invite people to this farm to feel the love of the Great Shepherd—for them to understand how much he loves them. If these precious friends know how much I love them, they might begin to sense how much they are loved by God. And even if they never do—even if the community they feel comes only from the people and the animals and the land surrounding them—it is still my honor and joy to love the people who are drawn to the farm. Every single person who sets foot on the place is someone I'm charged with caring about and respecting as a fellow human, created in God's image. I don't always accomplish this perfectly—sometimes I fall woefully short—but it is my goal.

In one of his last conversations with his disciples, Jesus asks his friend Peter, "Simon Peter, do you love me?" Peter responds, "Lord, you know I love you." "Then feed my sheep … care for my lambs."[1] Feeding Jesus' sheep—loving and caring for the people Jesus loves—is the task of Jesus' friends. It is also my task, the task that finally explains what I've been born to do.

CHAPTER THREE

The Paradigm: Stewardship vs. Ownership

steward |ˈst(y)oŏərd| noun a person employed to manage another's property, esp. a large house or estate.

a person whose responsibility it is to take care of something: farmers pride themselves on being stewards of the countryside.

Giving Away the Farm

The concept of "stewardship" forms the very basis of my understanding of my relationship to the farm.

A casual Tolkien nerd, I love the character of Denethor, the Steward of Gondor, in *The Lord of the Rings*. I love him as an example of how to be a Very Bad Steward. Denethor has been entrusted with the care and management of Gondor, the last great city of men in Middle Earth. He has made a royal mess of it. The once great and glorious jewel of a city has fallen into darkness, hopelessness and disorder. All of its riches and assets have been squandered. In the absence of the true King, Denethor has begun to think of the city as his own, and his greed and corruption have destroyed the very treasure he hoards. He has forgotten his place. He has given up the idea that the true King will one day return and call him to account for how he has managed his responsibility. And there is, almost literally, hell to pay.

A good steward, then, is the opposite of Denethor. A good steward simultaneously cares for her charge as if it were her own, tending it, nurturing it, sacrificing for it, growing and cultivating it, making it into more than it was when it was given, while at the same time living with the knowledge that the true owner will someday return and measure the steward's efforts.[2]

On the day that my family took possession of what would become the Jacob's Reward Farm property, we roamed all around the grounds. We looked over the ramshackle buildings, overgrown and broken-down fences, the wildflowers and weeds filling the pastures, and the beautiful, winding creek lined on both sides by willows and sycamores. "We own a creek!" I thought. I said it over and over in my head, trying on the concept. "We own a creek!" And then I laughed. It felt as foolish as saying, "we own those stars!" Regardless of what the property deed said, we didn't own that creek

The Paradigm: Stewardship vs. Ownership

any more than we owned the air we were breathing. But we had been entrusted with its care. We had been tasked with stewarding this property, to tame it and make it productive, in the service of the True King. I had no idea what that would look like. In fact, I'm still finding out what that should look like, every day. God is the Creator of this place, and I am the Steward. Not a bad gig.

I say this now, but this understanding did not come upon me all at once. My friend Laurie taught me the first critical lesson about the farm's real purpose. Laurie lives in a cookie-cutter suburb about 40 minutes west of me. One day a couple of months after we moved in, she called, fed up with her surroundings, and asked if she could come help me pull weeds in my garden. By then I had finally learned not to turn people down when they offered to help, so I gladly accepted. We spent the day digging in the dirt, planting, hoeing, laughing, watering, sharing stories of our lives, getting a little bit sunburned and turning that plot of soil into a productive garden. She called me later that evening to tell me how much the day had meant to her, that she had needed it so very badly and that it had refreshed her down to the depth of her soul. A little bell went off in my head. I wondered: if just a few hours of playing in the garden could have such a profound effect on someone, how much more potential was there for folks on this four and a half acres? The thought gelled into a sentence: "You're not the owner, Cindy; you're the Steward." This place wasn't for me to hoard, but for me to share. My perspective changed that day forever. I "owned" this farm for the benefit of others.

Stewarding is a huge privilege. Imagine someone handing you a massive hunk of gold and telling you to go make something remarkable out of it. The gold is shiny and pure, and the sheer volume sets you back on your heels. Sounds

exciting, right? Sure, first you need to take some goldsmithing lessons, and all the while you're learning, you plot and plan and dream about what exactly to make. You want to dream big enough so that the finished result *gobsmacks* everyone who sees it, but a part of you would rather soft-pedal the project in case your skills don't measure up to the potential. You might make a positive impact on hundreds, maybe thousands of people's lives, or you could end up looking like an idiot. You could even take the cowardly approach, risking nothing, and use the gold as your personal paperweight. Could you live with squandering such an immense resource out of laziness or fear? Not me. I risk looking like a fool every day, and sometimes succeed, but I choose to keep hammering on that gold nugget.

Stewarding a property like this farm means constant maintenance. It's a big job that never ends. First, there are the physical elements. In the spring, summer and fall, the grass and hedges need constant mowing and trimming. In the fall, winter and spring, building maintenance and capital improvements take center stage. Animal care happens every day regardless of weather, funds, energy reserves or the

The Paradigm: Stewardship vs. Ownership

farmer's mood. Then there are the interpersonal relationships that require tending, and finally, the vision casting and goal setting. We learn from mistakes as well as successes, and we try to do better next time. Stewards must plan and prepare for potential problems that might arise. We must dream big dreams and set big goals. We wear a hundred hats, from Manure Manager to Chief Financial Officer. We try to keep all the plates spinning, working hard to prevent any from falling and breaking.

The nice thing about being a Steward is that the ultimate responsibility is not mine. I don't have to know the future. I can't control the weather. I'm not responsible for the farm's ultimate end—only what I can see from my limited vantage point. This huge resource was bestowed on me, and the motive, strength and zeal to develop it seem to have come along in the package, by God's grace. Using the information I have, I've pointed toward goals, made periodic course corrections, and sometimes changed fundamental pieces of the plan to help the farm develop into its full potential. I'm at the helm, as it were, watching the stars and leaning on the rudder, steering into an unknown but exciting future.

CHAPTER FOUR

The fiber affair

I started designing knitted accessories around the same time that I started coming to the Barn. The Barnies have been my collaborators, contact makers, cheerleaders, testers and critiquers the whole time. Their support and encouragement keeps me coming back.

—LAURIE B., BARNIE

Giving Away the Farm

The adventure started back in the suburbs in 2001, when I learned to spin, learned to crochet and met my first shepherd mentors. It was a very big year.

My blessed mother-in-law, Julie, taught me to crochet. She was visiting at Christmas time, and I finally summoned the courage to ask if she would teach me. I had always been crafty and had come from a long line of crafters, but I was a generalist, with a couple of notable pauses for a ceramics or a cross-stitch phase. Julie had years of experience in most needlecraft skills—knitting, crochet, embroidery, sewing and quilting. She gardened, cooked and canned, and excelled at all the homey arts I loved. At my request, she and I went to the craft store and picked out acrylic yarn and a book. She spent a couple of evenings getting me started, and then she had to go back home. Left on my own, I kept plugging at crochet and taking to it pretty easily. Out of nowhere, yarn began to dominate every leisure moment. I birthed a full-blown obsession, in fact. My first crocheted piece was a very large baby blanket made for my newborn nephew, completely made up of single crochet stitches in a DK-weight yarn.

Those who know me know that I would never have been able to finish a large piece like that—which required the mind-numbing monotony of the most basic stitch repeated thousands of times—unless I was possessed with a trance-inducing compulsion. I couldn't put it down. The orderly rows and stitches, produced one after another, after another, fascinated me. I loved watching the fabric grow from manipulating some string with a stick. Crochet moved a lot faster than cross-stitch, which I found very satisfying. I could cover so much more real estate with a hook and yarn, that the investment of time felt much more worthwhile. Crochet is very architectural, too. You can "build" with it. When I teach new crocheters, I like to suggest to them that they think of

The Fiber Affair

each new row as a row of bricks—each stitch builds on the "brick" below it. Or, you can stick that hook in anywhere you please, and come up with fantastic freeform pieces that transcend rigid designs.

As much as I enjoyed it, that baby blanket nearly killed me. My right thumb almost quit working, and ached like a son-of-a-gun for weeks. But I was hooked (pardon the pun); I was a woman on a mission. I eventually built up some very specialized strength in that hand, and now I can go for hours and hours without cramping up at all. To this day, if I need the comfort of beautiful, soft yarn and mindless motion, I'll pick up a warm wooden hook and crochet nothing in particular. Or maybe the pragmatist in me will require that I can at least turn the work into a washcloth when I'm done. The joy of the craft is therapy, combined with the functionality of creating something beautiful and useful. From crochet, I learned the peace and comfort of yarn work for its own sake.

About the same time, while deep cleaning the drawers of my desk, I stumbled on an old brochure from the organization that has now become The Livestock Conservancy (www.livestockconservancy.org). This organization tracks and promotes the conservation of threatened heritage breeds of livestock—cows, horses, sheep, goats, domestic fowl and more. I remembered that I had tucked this literature away for the time I might be ready to own my own sheep. I can always be lured away from a deep cleaning job, so I snatched the opportunity for a little Googling—a new pastime for me in those days.

A little clickety-clickety on the keyboard and I was combing the web for Jacob sheep, my "someday" breed of choice because of their black and white spotted fleeces and their rare-breed status. Though they could be used as meat sheep, they were primarily a wool breed, which I preferred. I'm not

against eating lamb, but I wasn't sure I'd be able to raise animals that were intended for the table. How squeamish and softhearted would I be, come harvest time? I mean, after attending births, perhaps assisting with nursing, and watching lambs gambol through spring grass and generally peg the Darling Meter, I wasn't sure I'd be able to round them up and deliver them to the butcher. While trolling around, I stumbled on the fact that the president and the registrar of the Jacob Sheep Conservancy at that time were a married couple who lived twenty minutes from me. *Twenty minutes.* In all of the United States, the Big Cheeses of my chosen breed registry lived nearly close enough to hit with a rock.

Joan and Fred Horak lived in Lucas, Texas, on their 20-acre farm with more than a hundred Jacob sheep. They kindly invited my family out to visit their farm and talk at length about this wonderful sheep breed. I announced to them that they were my new best friends. For the next several years, the Horaks allowed me to hang out at their farm, pick their brains, help with chores, attend lambing and, most importantly, learn about wool. I learned to skirt fleeces (that is, pull the nastiest bits out of a newly shorn fleece), wash wool, hand-card wool, drum-card wool, and finally, spin wool. To this day, the smell of Orvus soap catapults me back to evenings washing batches of my first Jacob fleece in the kitchen sink. Goose-bumpy, woolly memories full of soap and lanolin and the unmistakable smell of sheep. I hand-carded that whole fleece into rolags and spun it while watching the 2002 Winter Olympics.

As the Registrar of the Jacob Sheep Conservancy, Joan told me that to belong to the JSC, I needed to own a Jacob sheep and have a farm name. She could see how badly I wanted to belong and to support the breed, so the Horaks sold me a sheep. Yes, of course I had to board it at their farm, but

Frosty was registered to me and that made me an Owner in the eyes of the JSC. Then, I needed a farm name; nobody said you had to have an actual farm. Back I went to Google to find some reference to Jacob that wasn't already in use—I just felt like there should be some play on words with the name of the farm and the breed. Jacob's Ladder? Nope. Jacob's Fold? Nope. I recalled the story of Jacob in the Old Testament (the breed's namesake). Jacob had plenty of attributes, and many adventures. His was the very first recorded instance of a colored wool breeding program.[3] He worked very hard and suffered a good deal of injustice, but eventually became a very successful man. I settled on "Jacob's Reward" for my farm name. Still stuck in the 'burbs, but a sheep owner and a member of a breed association, I shared determination and stubbornness with the Patriarch.

And I bided my time.

Hanging around all these sheep and fleeces, the next logical question occurred to me: what's stopping me from (gasp) learning to spin? It seemed crazy, but then, why not? I had located Jacob sheep with no trouble at all. Perhaps I could find a spinning teacher within a reasonable distance. After several days of research and half a dozen phone calls

(spinners weren't that adept at using the internet yet), I found a very talented lady, skilled in many fiber arts, who was offering a class. She also raised Corriedale sheep and Angora goats, so I learned even more about working with fiber than before. I fell helplessly in love with wool and all fibers. This was worse (or at least as bad) as any high school crush, because I have never recovered.

I was going to need a spinning wheel, and soon. As I plotted and planned how I would justify the expense and the wacky new apparatus occupying floor space in the house, a little bell went off in my head. My sister-in-law had at one time been a spinner and a weaver. She had learned to spin when she and her Air Force officer husband had been stationed in Colorado, and she still owned the loom and the wheel. She hadn't spun in a long time, but she told me she was planning on taking her wheel with her to their next assignment in Germany. Drat—I'd have to buy my own. What could I do? I bit the bullet and bought the Ashford Traveler I had learned on, and my life as a committed spinner began. I had the power to turn smelly sheep hair into fabulous yarn. Textile alchemy.

The next year during her Christmas visit, my mother-in-law helped me add knitting to my yarn repertoire. Two needles with lots of live stitches increased the degree of difficulty a bit, but once I mastered the basics, and relaxed into the rhythm of knitting, a whole new world of stitchery opened up. I couldn't get enough of yarn, so I was anxious to conquer any new skill that kept me close to the woolly stuff.

The downside to all this newfound yarn knowledge and sheepy resources was that I was still stuck in the suburbs, with no sign of that changing. I had no idea if or when I'd ever be a real shepherd myself. I had to be content to enjoy ovine pleasures through visits to the Horaks' farm and through knitting, spinning and crochet.

My gawky new skills found a home in a yarn-crafting group that met weekly about a half hour from my house. While my daughter was in school, I slipped off on Wednesdays to hang out with my new friends. These ladies, most of them retired empty nesters, had a lot of spinning and stitching experience to share with this newbie, and they taught me plenty of skills. We met every week and shared our lives, our finished and unfinished projects, our triumphs and tragedies. I began to see how this love of stitching and wool and beautiful tools and color and texture could meld a group of artists into a group who needed and nourished each other. We were all hungry to learn everything we could and happy to share the secrets we discovered. This group also taught me the value of creative synergy and planted lots of good seeds that would sprout and grow in due time. Thirteen years later, these ladies are still an important part of my beloved community. Little did I realize how much that community was about to grow.

CHAPTER FIVE

At Last, a Shepherd

*What I found at the farm was a community unlike any other knitting group I'd ever sat in with. This group was so friendly and inviting and **loving**. It felt like a warm hug—a knitting home I didn't even know I was looking for.*

—Lisa, Barnie

Giving Away the Farm

On the day before Mother's Day in 2004, I coerced my family into driving over to a friend's house in a neighboring town so that I could spin with a group of ladies she was hosting in her living room. After all, it *was* almost Mother's Day. I figured my husband and daughter could hang out with my friend's family for a bit and visit the alpacas they raised on their small acreage. The promise of seeing fuzzy alpacas up close did the trick. I enjoyed a couple of hours of spinning and chatting with my friend and a small handful of acquaintances, while Ted and Emma experienced their first up-close encounter with the small, fluffy camelids. Inside, the topic of land came up in conversation—specifically, when was I ever going to find a place to have the little farm I had always wanted? My friend happened to mention in passing that there was a piece of property just around the corner for sale: 4.5 acres with a house and some outbuildings.

Really? Again, I imposed on my family—this time to swing by the place on our way home.

It was beautiful! Big pastures, tall trees, a creek, with barns and sheds already in place. The house looked small, but it was worth investigating. I took down the realtor's phone number and placed the call on the spot. He told me the price—unbelievably low—but warned that I shouldn't get too excited. The property was in the flood plain, and the house was pretty much worthless inside. My euphoria was dampened, but only a little.

And thus began two months of research and due diligence. I learned a lot about flood plain maps, building codes and mortgage requirements. After satisfying my husband's concerns about whether the place could be rehabilitated, and under a healthy layer of denial and naiveté, we cobbled together some unconventional financing and bought the place. We launched into our new roles as landowners and

At Last, a Shepherd

general contractors, trying not to notice all the parallels between our situation and the plot of the '80s cinematic hit *The Money Pit*. An entire book could be written about our eleven months of demolishing and rebuilding, but suffice it to say that the construction adventure finally ended and we moved in on July 1, 2005. Several months later, our house in the suburbs sold, and the migration was complete. We were officially country folk.

Not far from the house stood a dilapidated 24'x12' storage shed, built to look like a barn, covered in red paint that had oxidized to a powdery pink and bearing big, swinging doors with white trim. The window in the rear was shattered, and inside, arranged as though by a gentle tornado, lay all manner of broken things—most of a ceiling fan, car parts, raccoon droppings, cobwebs, bits of toys, boxes and sundry workings of entropy. This was going to take some elbow grease, but I already had designs on it as my studio.

Giving Away the Farm

In the flood plain, you are not allowed to tear down a building and rebuild it. All available unpaved land is needed to absorb floodwaters, so if you tear anything down, FEMA reclaims that land to help reduce flooding. You're only allowed to perform "maintenance" on existing structures grandfathered in on the property maps. So we "maintained" that red barn. We "maintained" it to within an inch of its life. It got a new foundation, which lifted it up out of the dirt. It got new siding, new framing, a new roof, a wonderful new cupola with a sheep weather vane, new wiring and sheetrock, tough Berber carpet, built-in shelving and a combination AC/heating unit. We were going to be comfy year round in there. Finally I would have room for all my fiber stuff. The Little Red Barn was born.

As we were still settling into country life, I got a call from Joan Horak. My second boarded Jacob sheep, Israel, had turned up lame, and the other sheep were running him ragged, aggravating the situation. She asked if we had a place we could keep him so he could recuperate in peace. "You mean, bring him here? For good?" I asked. Yes, she replied, she thought it was time for Israel to come to his new home. I panicked. Here I was on the threshold of shepherd-hood and I didn't feel ready. I felt grossly unprepared. Sure, I had read all the books and had stuffed my head with reams of information, but now the reality was coming at me like a train and I was scared.

Was I ready to be the Perfect Shepherd? (For that, of course, was my unconscious goal.) That was why we had moved out here, right? Ted and I raced outside and quickly cobbled together a pen around the one structure on the place that had once held animals: a ramshackle, two-stall wooden shed that leaned at about 20 degrees. I hoped it wouldn't fall down in a stiff wind. Joan and Fred pulled up with their shiny

red trailer and unloaded our first livestock. We turned Israel into the pen, and the Horaks drove away.

I was a shepherd.

Israel cried for a week, because sheep don't like to live alone. They are flock animals and only feel safe and secure in groups. I faced my first shepherd crisis. I called the Horaks and asked if they had any other sheep I could buy to keep Israel company. (I had, unfortunately, sold Frosty several months previously.) They didn't have any for sale, but they decided to give me one of their older wethers, a neutered male named Tommy. Tommy was bottle-raised and very personable. He was the sheep the Horaks would introduce to visitors because he would happily come to the fence for scritches while the other members of the flock stood off at a distance, aloof. I couldn't believe Tommy was coming to my farm to live. The two sheep bonded pretty well and seemed

content in the little pen and the wobbly shed. These two sheep began to teach me how to translate book knowledge into practical application, and I began building the daily rhythm and routine of a farmer into my new life in the country.

Little did I know, the snowball of livestock ownership was rolling downhill and picking up speed.

I had barely gotten my bearings as a shepherd, when my friend Cyndi up the road called to tell me she had a couple of alpacas for sale. More accurately, the gal who boarded alpacas at Cyndi's ranch was selling a couple of her males. Was I interested?

Alpacas? I had never even considered it. I assumed that they were crazy expensive and probably fussy to care for. The fiber was wonderful, of course, but … alpacas? What was a girl to do but jump in the car and tear over there immediately to see them for herself? I felt myself hurtling toward the inevitable.

Sure enough, the two male alpacas in question were beautiful in my eyes. They weren't showstoppers, but they were fairly tame and beautifully fleeced. One was a dark fawn color, the other a mahogany brown. She wanted $500 each for them—a fraction of what I expected to pay, if TV infomercials were to be believed. And I'd have to take both of them, because like other herd animals, alpacas don't do well without at least one other of their own kind. But I didn't have $1000 laying around, so I had to put on my thinking cap. How could I 1) come up with that kind of cash, 2) change my whole business plan, and 3) ALPACAS! Then it hit me: I could get someone else to buy the second alpaca and board it at my farm.

It worked. My friend Barbara had been jonesing for the alpaca lifestyle, but lived in town. I called her and laid out the crazy plan, and she jumped at it. Her long-range plans involved moving to the country and starting her own farm, so this would help her begin her dream until she could actually get onto her own property. I would buy Moonstruck, the young fawn boy, and she would buy Gizmo, the older brown male, and pay me a small amount per day to take care of him. I put together a revenue projection proposal to convince my husband that this was a good idea, mostly based on fantasy, and then all I had to do was come up with $500 out of thin air. Easy peasy. I convinced my friend Marlene to buy the Bernina sewing machine and serger I had recently inherited. (My dear departed Aunt Jo would have been totally OK with it, I have no doubt.)

And the snowball grew. We added donkeys to guard the alpacas, chickens for eggs and more sheep for wool. As the oldest of three sisters, I had carried the scourge of hyper-responsibility through my life, and I easily applied that tendency to my new role as shepherd and farmer. Should

Giving Away the Farm

I put coats on the donkeys? They looked so cold! Did the chicken coop need a heater? How often should I change the water in the troughs? At first I actually had a hard time leaving the property for fear something would happen to one of the animals in my absence. Eventually, I was able to relax a bit, coaching myself with logic and reason: "Yes, farmers are quite able to go to the grocery store from time to time and nobody dies." I learned how to stack and toss hay, how to give shots and take rectal temperatures, trim toenails, check weights, administer medicine and more. I fussed and overprotected, but with the exception of some suicidal chickens, everybody survived my learning curve.

Almost. Our first serious livestock tragedy came right after Christmas in 2009. My little sheep flock was now made up of a motley collection of Jacobs, Gulf Coast Native sheep, a Suffolk cross, a "fine wool cross" sheep I bought from an FFA student and a Cormo cross ewe I had gotten from Animal Control. Trusting the advertising, I had them enclosed in special electric fencing up on the north end of the property where there was as yet no tall, welded fencing to keep them contained. All went well for months and months. But two days after Christmas, Ted raced in from an early walk with the dog, shaking me awake and telling me I needed to come out right away—coyotes were attacking the sheep. I've never dressed so quickly in my life. We dashed back outside to find the flock cowering in terror, coyotes long gone, and two of my sheep torn to bits. The smallest little Gulf Coast sheep, Eli, was dead, and so was Queen Esther, the Cormo ewe. I was devastated. I had failed my flock. It was a Sunday morning, and I could hardly make it through church that chilly day, still singing about the Christmas shepherds watching their flocks by night. Despair turned to anger and determination. I was getting a guard dog.

Up to that point, I had been ambivalent about livestock guardian dogs. The most popular breed around here is the Great Pyrenees, but they seemed to have some specific downsides: they are reported to need a very large area to guard, and if they don't have it, they are likely to jump the fences and patrol what they believe is their territory. They can be bad about both jumping fences and digging under them. Some Pyrs are great with sheep but kill chickens. Success or failure depends mostly on the individual dog. But this situation with the coyotes and the sheep steeled me to face any challenges that might arise with a good guard dog. I contacted the local Great Pyrenees rescue group and researched the options.

Even rescue dogs come with a price tag, and the adoption fee for a healthy, neutered, evaluated livestock guardian dog was $300—again, money that just wasn't in the budget. While I was casting about looking for something to sell to raise the cash, my Wednesday spinning group rallied around me. Secretly, they passed a hat and came up with every penny I needed to get the dog. I wept for joy and gratitude.

Judah came to the farm in late January and settled in immediately. All my fears about the breed vanished. He acted like he had always lived here, calmly patrolling the perimeter and staying between "his" sheep and any perceived danger. The sheep took a little longer to warm up to him. But in almost six years, Judah has never once tried to leave his paddock. He is a bit overprotective of me, so visits to the vet can be challenging, but while on the job, his work has been impeccable. Now between our tight, woven-wire fence and dog protection, our sheep may safely graze.

CHAPTER SIX

The Siren Call of Yarn and Critters

I have a hard time fitting in [elsewhere] because I am an introvert. I seem to be able to fit in with this group rather well. I am accepted for who I am.

—Denise, Barnie

Giving Away the Farm

As you first drive up to Jacob's Reward Farm, the first thing you appreciate about it is the space. The sheer openness of the landscape. That's one of the things that folks from the 'burbs are craving: the ability to swing their arms in any direction and not hit a Starbucks or a filling station. The city of Parker has a two-acre minimum lot size, so most of the neighborhoods enjoy this open, spacious feel. When people drive out here, they step out of their cars and visibly take a deep breath. Space is like gold. It's what I have been longing for most of my life and what drove me to a rural life in the first place. I assumed we'd figure out what to do with the space once we had it. Garden, sheep pasture, polo field—who knew? But we'd cross that bridge when we got to it.

Once we moved to the farm in Parker and began shouldering the work required to keep the property and livestock managed, I found it harder and harder to continue traveling 30 minutes each way to my spinning group every week. I hated the idea of giving up time with my friends, but the demands of the farm never took a day off. I had to make different choices now that I owned animals and land. Even my vegetable garden suffered more and more neglect as my livestock populations increased. Something always has to give. There are only 24 hours in a day, after all, and each day's To Do list perpetually spilled into tomorrow. I began slowly redesigning my life around the amenities to be found here on the property so I didn't have to leave. I particularly wanted to utilize the Little Red Barn, my snug little Hobbit hole of fibery treasures. On the third Saturday of each month, I opened up the place to any knitters, spinners, crocheters and other fiber artists to come visit. We could comfortably seat up to 14 people in the room, with space for spinning wheels when necessary. Laurie brought her signature quiche most months, and others brought snacks to share as well. I put on a pot of coffee, and we were cozy and comfortable.

Our Saturday crowd grew. To keep communication flowing, I started a Ravelry group, a MeetUp group, a Facebook group, and had a listing on Local Harvest. I wanted all local people within the sound of my virtual voice to enjoy the magic of the farm. Pretty soon, our get-togethers drew almost more people than we could fit into the LRB (as we had affectionately dubbed the Little Red Barn). We kidded about enlarging the building, but since that was not really an option, I decided to "enlarge" the month. We added the fourth Saturday to the monthly calendar for getting together. And still they came.

Here's the miraculous bit: we've documented a phenomenon we've named "the Economy of the LRB," where there is always a seat for anyone who shows up to occupy it. Some people come early and leave early; others come later and hang around to the bitter end; and always, there is an empty chair. When we think we cannot accommodate another knitter, a seat opens up—and not from someone feeling pressure to leave. It just happens. Everyone who is supposed to be there, gets there. We're constantly amazed.

While the volume of people drawn to the LRB surprised me, the extreme diversity of the group was a greater surprise. Men and women, young and old, many faiths or none, homemakers and professionals, a spectrum of political persuasions, and every conceivable yarn and fiber craft have all been represented, sometimes in one day. Under certain circumstances, this could be a recipe for disaster, but we have always managed to get along and have an uproariously good time. A few of our "Barnies" had attended other groups, but didn't feel as comfortable as they did in the LRB. Some folks happily attend more than one group in town. For some others, this was their first craft group experience.

Giving Away the Farm

But from each occupant of the LRB, the response has been the same: all feel welcome, accepted and valued here. Not everyone who has ever visited has stuck. Some folks have moved on, and for lots of reasons. Some have taken short sabbaticals and then returned. We can't possibly be the best fit for everyone, but to my knowledge no one has left mad or hurt. How does this happen? Why are there no cliques or disputes or splits? I'll discuss some theories in a bit.

I admit our particular location is unique, with its wide-open spaces, the animals, the fresh air and refreshing proximity to nature. As a result, we usually host people with one of at least three different interests. First, the hard-core yarn crafters (knitters, crocheters, spinners, weavers, felters, etc.) come to the LRB on the third and fourth Saturdays and don't spend much time in the pastures. They greet the occasional dog or cat or chicken that might stroll through the LRB, but they're not in any rush to get outside. Some are actually a little put off by all that "nature" and come out to the farm in spite of the wasps and the port-a-potty.

The second group loves the animals and insists on greeting the guard dogs or the alpacas or Shadrach the sheep before they settle into the LRB for some social time. Some of these folks can be counted on to help out with chores in a pinch. This group also includes the home school groups, the Scout troops, the daycare centers, and families who arrange for special tours of the farm, all interested in learning more about the animals and farm life. We also do wool processing and spinning demonstrations for tour groups, and everyone gets to wear home the wool bracelets that they spun themselves. I've done my job if everyone leaves the farm with a new awe of and appreciation for wool and its magical qualities.

The third group of people attracted to the farm lives where the Venn diagram overlaps—love for fiber *and* the

farm. They may harbor dreams of one day owning a fiber farm very much like Jacob's Reward, and they are always on the lookout for some experience or knowledge that will one day help them realize their dream. For these farm friends, we offer workshops and volunteer opportunities, and will be developing consulting and online learning programs to mentor them through the process. I've been in their shoes, and I appreciate every scrap of knowledge anyone offered me while I was learning to be a shepherd. I'm very anxious to turn around and offer all my sister and brother shepherds a helping hand.

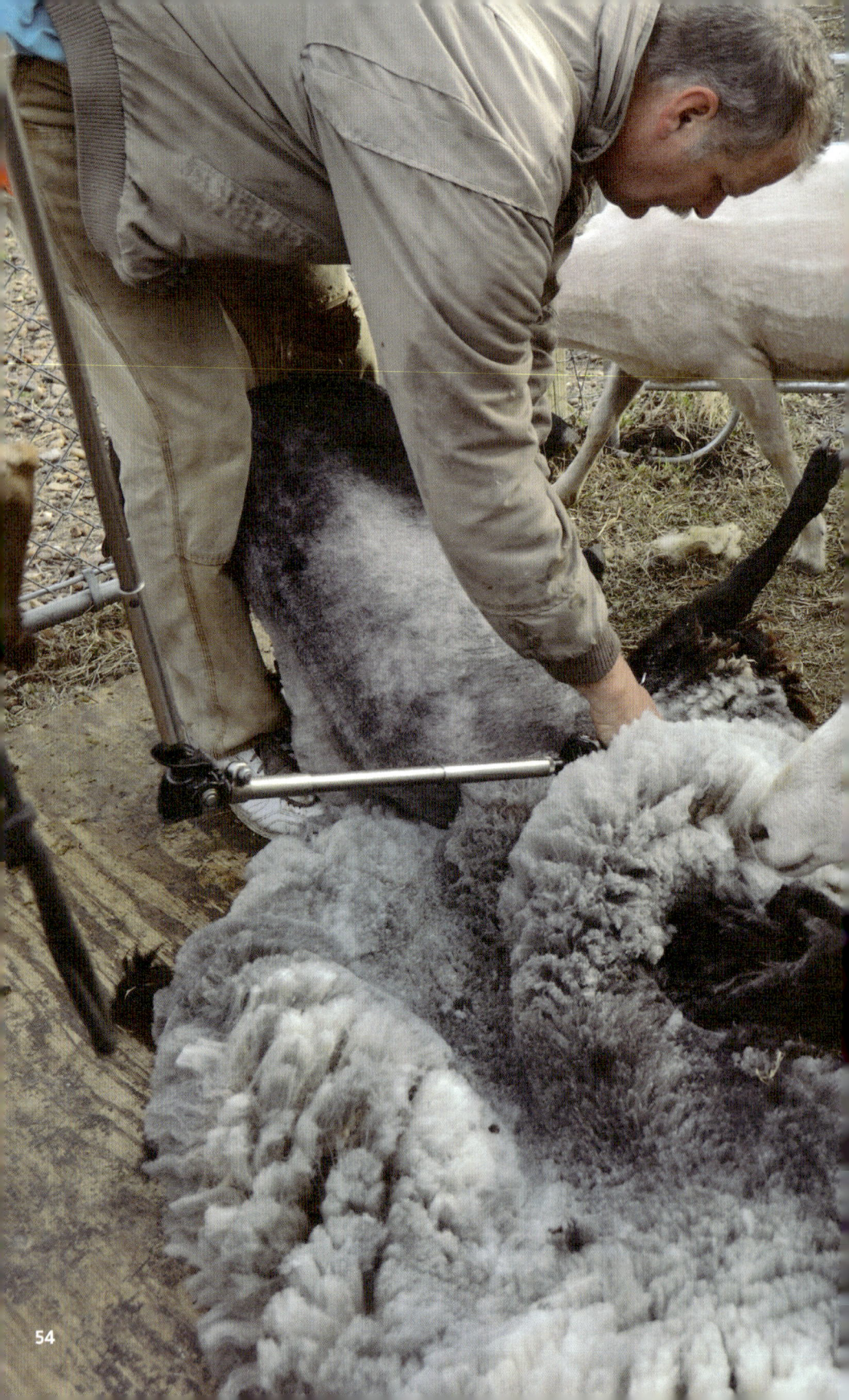

CHAPTER SEVEN

The fiber CSA: A Better Way

My first actual visit was after I'd had a fiber-y holiday in New Zealand. I didn't want to let go of that country community feeling I'd got in a place halfway around the world. YEA—here was a little spot even better, and close to home, too.

—Jane, Barnie

Giving Away the Farm

All this Saturday knitting and earthy Kumbaya is great, but a farmer's got to bring in some revenue. I looked all over for some innovative ideas about how the farm could support itself.

Through the knitting and crochet website Ravelry.com, I met Susie Gibbs, a shepherd who at the time lived in New York's Hudson Valley. Susie is an engaging writer and talented photographer who had gathered quite a community around her farm. Her warm writing style and gorgeous photos drew me to check out how she was running her business. The most remarkable thing she had done was to apply a unique business model—the CSA (which stands for Community Supported Agriculture)—to her wool and mohair "crops." CSAs offer "shareholders" a proportional share of the harvest in exchange for an up-front investment. In the case of fiber, shareholders pay a certain amount in advance, and

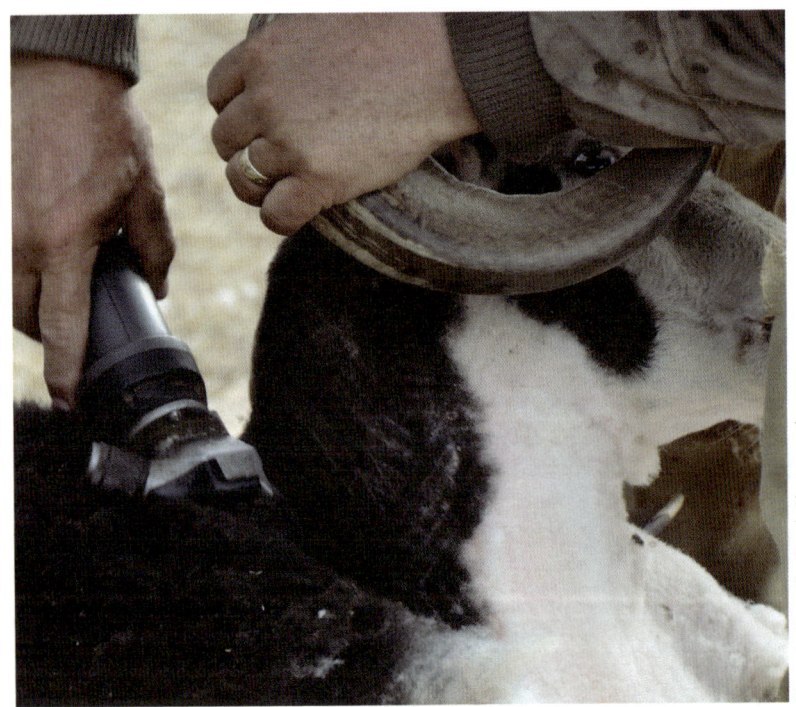

Photo by Jennifer Jurek Photography

then when the wool is sheared and processed, the harvest (or "clip") is divided between them. Some fiber CSAs distribute wool that is ready for spinning, while others distribute spun yarn ready for knitting—and some offer a combination of both.

Not only does this model inject a nice chunk of capital into a farm all at one time, but it begins the process of developing lifelong friends and farm supporters. Knitters and spinners who invest their money also invest their hearts and attention. They learn about the individual animals whose fiber they will soon be knitting or spinning. They develop their favorites, and if a farmer is smart, she will keep these animals' stories front and center in her communications about the farm. Several of my alpacas and sheep have their own informal fan clubs!

Fortunately for me, Susie grew up about an hour from me, in Fort Worth, and her mom still lives there. I learned that Susie was coming home for a visit, so I finagled a meeting with her to learn all I possibly could about her CSA, her farm operation and her loyal community. We got very large cups of coffee at a Barnes and Noble bookstore, and for a couple of hours Susie proceeded to pour out her valuable business tips. I scribbled everything down as fast as I could—every word was a gem. Susie's generosity with her business ideas overwhelmed me, and I will always be grateful. I've used those ideas for years, and they became the springboard for many more.

Not only was she generous with me by sharing all her business "secrets," but those secrets revealed her consistent generosity toward her customers and community. Her business grew precisely because she gave away so much of herself, day in and day out. Perhaps most dramatically, Susie gave away several small flocks of sheep and goats to aspiring fiber farmers after a rigorous process of vetting the

applicants for suitability. She cared about the new farmers and about her animals, and she took the time to make sure each had a good chance of success with the other. She asked the applicants to submit essays or videos about why they wanted to win the flocks, and she chose the winners with the help of her community members. Susie involves her community in all her projects, not only strengthening their bonds, but also benefitting from their input. I have learned a great deal from watching Susie's transparency in business and the synergy with her ever-growing fan base.

At the Maryland Sheep and Wool Festival several years ago, I had the opportunity to hear Susie speak about her business—which by then had moved to Virginia and taken on the name Juniper Moon Farm. She stood up to speak right after a gentleman who did not exactly share her confidence in the strength of generosity. He had exhorted the audience to do whatever they could to capture customers' business, because "it's a zero-sum game, and if they buy from someone else, they won't buy from you."

When it was Susie's turn to speak, she immediately turned the tables: "With all due respect, I have to disagree with the gentleman who just spoke. I don't believe it is a zero-sum game. One person's success doesn't have to come at the expense of the next person. There will always be yarn buyers, and they will always find room for one more skein if it's a good product and if you're nice to them." Or something to that effect. Susie's words made perfect sense to me, both as a knitter and as a business person. The view that "the universe is stingy; you have to look out for Number One" only leads to stinginess in business. It leads to business owners who treat their customers like vending machines, always trying to get the most out of them with the least investment of product or service. Customers object to this kind of treatment and will likely go elsewhere for what they need. On the other hand, the idea that there's plenty of just about everything to go around gives us permission to lavishly bestow on our customers service, value, time and respect. That kind of treatment encourages our customers to come back to us again and again.

In related news, treating our *friends* generously keeps them coming back as well. Givers have lots of friends. Takers, not so much.

Susie has gone on to develop a commercial yarn line, and other creative projects with knitting designers, and she's less involved with her flock day to day, but her connection with her community is strong. As a result of these years of investment in them, she has a reliable body of loyal friends in her corner for whatever amazing projects she launches. Her investment and generosity will be paying off for many years to come.

Photo by Jennifer Jurek Photography

CHAPTER EIGHT

The Must-have Conditions: Hospitality, Acceptance and Compassion

What I love the most is that everyone—everyone—checks his or her attitude at the door. I've not encountered any Queen Bees or Alpha Knitters pushing this group around; Miss Cindy keeps a firm but gentle hand on her party, and the result is truly magical.

—Lisa, Barnie

Giving Away the Farm

When newcomers visit the farm, during the LRB days in particular, they report feeling welcome, relaxed and at home. An atmosphere pervades that encourages them to feel a part of the group pretty quickly. This culture of the Barn developed very organically, without rules or by-laws or even spoken intentions. So it took some deeper investigation to sift out what accounted for the friendly, welcoming spirit. When I polled the Barnies to see what had drawn them here and then kept them here, several themes emerged.

The Barn Culture Is Hospitable

I was definitely standing behind the door when God gave out the gift of hospitality. My mom, in contrast, had been trained as a young lady in the '50s to elevate hostessing to a fine art. Once, while still in high school, she entertained some "gentlemen callers" at her home on a Sunday afternoon. She dressed up, served crustless sandwiches with unrivaled aplomb, and her soirée ended up as a brief news item in her local paper. It might have been a slow news day, but achievement in the social graces was celebrated and rewarded in those days. She continued the tradition as a military wife, hostessing luncheons and cocktail parties when I was a little kid. She always did such a great job that I found it simpler to just stay out of the way while she worked her magic. I grew up feeling distinctly unsophisticated. This is of little consequence when you're a kid. But eventually, as an adult, I needed to learn some basic etiquette and empathy so that I could anticipate people's fundamental needs, like food and drink and directions to the restroom.

I don't remember the exact moment it happened, but at some point early in the life of the LRB, I decided to quit

The Must-Have Heart Conditions: Hospitality, Acceptance and Compassion

rebelling against the pressure to be like my mom, and relaxed into a more casual style of friendliness and welcoming. I made the process more my own, reflecting the more casual attitude of my own times and my own personality. I have worked intentionally to find ways to help guests feel comfortable in my home and here at the farm. But I'm a lightweight compared to some of the Barnies.

Take, for example, our friend Laurie. Nothing says lovin' like something from the oven, and as I have mentioned, Laurie is a great cook. She loves people unconditionally and loves to show love with food. She hardly ever shows up to the farm without some delectable main dish or healthy snacks to share. We've found that eating together bonds the group and makes the gathering into a party, and so noshing is a wonderful part of our Saturdays. It also is a great excuse to put our needles down or our spindles aside every once in a while. It gives our newcomers a great way to meld into the group in short order.

If old-school hospitality is still alive and well in the fellowship that food brings, I have also learned the ongoing value of learning and remembering visitors' names, because our names are important to us. When people know our names, they're half way down the path to knowing us. This is a learned skill for me, so I have tried to make a habit of going around the circle when we have visitors to introduce everyone to the group. It's a fun way of pointing out something memorable about each person, letting the newcomers in on silly inside jokes and calling out everyone's unique contribution to the group. Nobody leaves feeling overlooked or ignored. By the end of a visitor's first day with us, they are one of the pack. Even those who don't return remember the farm fondly as a place where people cared about them.

We welcome each other as fiber artists, too. Newcomers are invited to show off their work, and the Barnies respond with appropriate praise. They always find beauty in the visitors' pieces, sincerely expressing appreciation for the skill we all can recognize. The Barnies instinctively gather the new person into the fold as "one of us." Yes, I'll say it: yarn knits friendships together.

Beyond the basics of hospitality, there's an even deeper quality that sets the stage for great community: acceptance.

The Barn Culture Is Accepting and Respectful

Left to our own devices, most of us tend to travel in pretty homogenous circles, hanging with folks who look like we look, and who believe what we believe. We gravitate to folks who are like us in culture, economic circumstance, political bent, fashion style, values, or education—that's natural and not intrinsically bad. But it limits our ability to think outside our own boxes or see the world through different eyes. We tend to begin fashioning an environment finitely described by our own experience and expectations. When confronted with different perspectives or cultures, we may react with unflattering defensiveness or caginess, or even obvious unkindness.

Lots of circumstances built into my life the idea of compassion for my fellow human beings. Of course, my Christian faith encouraged me to "love my neighbors"—those people I liked and also those people I didn't like—but you can go for a long time without having to actually love people who are very different from yourself. What does it really mean to "love my neighbor" anyway? I doubt that when God issued that commandment that He meant "wave to the family next door when you pass them in the alley twice a year," or "try

The Must-Have Heart Conditions: Hospitality, Acceptance and Compassion

hard not to flip off that annoying guy in traffic on your way to work." No, I think the message was more like: "Sacrifice. Do for others what you think it would be awesome for someone to do for you, even if it's inconvenient or costly." That lofty concept can live up on a mental shelf for years without having to be taken down, dusted off and used. As long as we limit ourselves to homogenous society, we won't have our ideas challenged or our sensibilities offended. It's easier, but not very rich.

I'm grateful for my years in various occupations dealing with the public and then building a business of my own. Those years taught me a lot about how to build bridges of commonality with new friends, allowing them to become old friends.

The Barn is a place where many different kinds of people come together. And maybe this is the place where I have

intentionally stood up to model kindness and tolerance, in the classic meaning of the word. The relationships that are so important to us must be guarded and nurtured. It takes time, energy and intention to protect the cohesion of a diverse community. Friendship must not be conditional—we must commit to being friends even if we disagree on the most foundational issues, because we care first and foremost about each other's good and about truth. We need to work hard to get along with people we don't want to lose. If there is an unspoken rule at the LRB, it is this: We will always show each other respect and kindness, no matter what. We make room for differing opinions and personal styles, but there is no room for slander or rudeness. Ever. If you come to my Barn, I will defend your dignity against all comers.

Don't let me give you the wrong idea—there have been no dramatic instances of loud discussions about this. It's simply our culture. I have occasionally gently steered the conversation in another direction if we seem to be heading down a dangerous path, or I may change the subject altogether in a lighthearted way. I try to be subtle, but the Barnies tell me that my intentions are clear. We will show respect to each other.

Why does it take practice for us to treat each other with the respect we deserve? Because most of us, myself included, get wrapped up in our own lives and priorities, and we can forget to look our friends and neighbors in the eye to really see them. It takes energy. It means slowing down. It means setting aside our own agenda for the moment. But it can become a way of life. Even in life away from the farm, I began several years ago challenging myself to always look store clerks and waitresses in the eye, and warmly and sincerely thank them for their help. If they have a nametag on, I call them by name. This isn't just a courtesy. It comes out of a deep recognition that each person

The Must-Have Heart Conditions: Hospitality, Acceptance and Compassion

is an extremely valuable member of my human family, with a future and a history and the potential for greatness. And the wide-eyed, grinning responses of these people encourage me to keep reaching out. What I've learned is that everyone I meet has a deep need to be acknowledged and loved. When they receive even the slightest show of thanks or courtesy, they often light up. Because unfortunately, we live in a world where common courtesy seems to be scarce. It costs so little to smile and be kind, and it pays off. I'm not doing the other person a favor by acknowledging their worth. When he or she smiles back, I am the one who wins. And wouldn't it be great to be known as someone who leaves a trail of warmth and encouragement everywhere you go?

The Barn Culture Is Compassionate

All relationships cause us to grow and change. Committed friends can smooth off each other's rough edges as they interact over time. Barnies regularly take the opportunity to grow and change as they invest in the friendships they find at the LRB. I've watched uptight people soften and make allowances for strong personalities they might not have tolerated before. I've watched them make the mental choice to hold their tongues rather than spout criticism. It's amazing to watch, and a privilege. As we invest in each other, trust grows. As trust grows, vulnerability deepens. When we're vulnerable together, we can love each other better, and help each other heal from the little nicks and cuts we have suffered in the "outside world."

Where can you go when the world seems to have painted a big red target on your backside? You got it: the Little Red Barn.

One of our Barnie pals—let's call her Marilyn—texted me one Saturday morning to ask if she could just come get a

Giving Away the Farm

little peace and quiet in the LRB, knowing it wasn't one of our regular get-together Saturdays. She had spent a hard week vacationing with family, and that had taken quite a toll on her. What had been billed as a time for some siblings to get together and enjoy each other's company turned into a marathon group backbiting event. Ouch. That called for some extended Barn Time.

Marilyn showed up with her knitting projects, and I left her alone with her yarn and her thoughts while I went about my morning farm chores. When I finished, I joined her in the barn with a big cup of coffee for each of us and took up my own stitching. I gratefully accepted the excuse to sit and crochet while Marilyn poured out her painful story.

I was so humbled to see that she felt safe enough here to let it all hang out—to air the litany of petty incidents that had ruined her time away, all the accumulated, tiny cruelties that had stolen her spare thoughts even after her return. As the hours slipped smoothly by, we shared lots and lots of stories of our lives. I learned so much about my friend that our previous visits had not revealed. I got to see lines of tension melt off her face and sweet, relaxed smiles replace them. We laughed. I'm not sure she had laughed in a while.

I even got to sit still long enough to get some spinning done! Marilyn was shocked and glad that her visit had given me just the excuse I needed to carve out a bit of "me time," too.

This, I believe, begins to approach the core of what makes our community special. We're here for each other. What may have begun as a love of knitting, yarn, or fluffy critters, blooms and grows into a love for the other people similarly attracted. We're drawn together by the stuff of the farm: the open air, the smell of fresh dirt, the doe eyes of the alpacas, the silky softness of the shorn wool, the hot coffee on a brisk day—who knows what element sums it up for each person—

but we find ourselves gathered in the cozy barn every third or fourth Saturday, anxious to pick up where we left off last time—anxious to be known and valued a little more deeply.

The Barn Culture Has Team Spirit

I love swag. I love wearing the colors and logos of groups to which I belong because it instantly identifies me as One of Them. So at the LRB, we have lots of swag. We have embroidered ball caps, aprons, t-shirts and jackets. We have luxury handcrafted drop spindles made by Mr. Tom Golding[4] with the farm logo on it. Soon, we'll have sterling silver charms for necklaces or bracelets. We have not yet run out of ways to show whose team we're on.

CHAPTER NINE

The Most Unlikely Leader

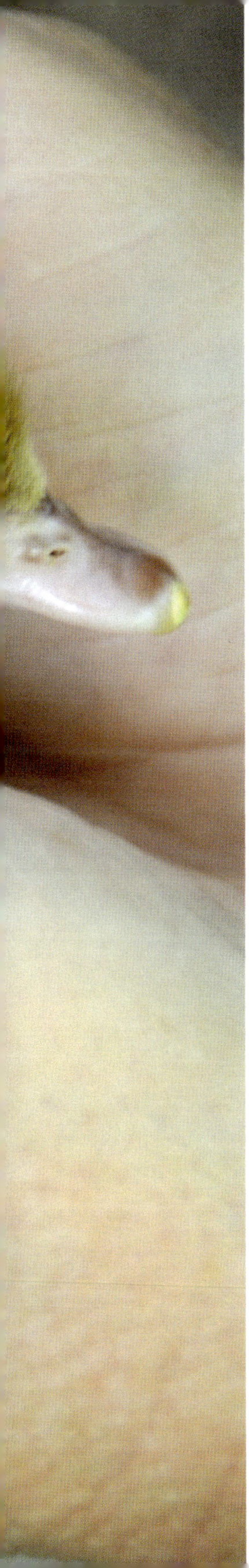

We have had many different adventures, from wrangling and shearing alpacas, to building fences, to breeding rabbits, to dyeing fiber. Going to the farm lights up my day, and when I come home, the stories I get to tell David of the adventures make him smile. He can see that it truly makes me happy. On the farm, I feel like I am doing some good for the world.

—Victoria, Barnie

Photo by Jennifer Jurek Photography

Introvert vs. Extrovert

I'm not a trained psychologist, but my understanding of introversion and extroversion is this: extroverts are energized by interaction with other people and introverts are energized by time alone. Each gets drained by the trials and stresses of life and must regularly visit their source of refreshment and strength—solitude for introverts, and company for extroverts. Extroverts can be quiet and thoughtful when they need to be, and introverts can be outgoing and gregarious for short spurts, but when the energy is used up they both need to recharge in their own unique ways.

Are most farmers introverts? Perhaps farming attracts introverts because of the isolated, rural lifestyle they pursue. We think of the old farmer out on his tractor for hours, or stuck in the milking parlor twice a day or rising before dawn while his family sleeps. Young couples yearning for the "back-to-nature" homestead may be running from the hustle-bustle of city or suburban life, and need to find lots of open space and empty sky. When I think of farmers, I think of lone souls shoveling manure, herding livestock, weeding gardens and birthing babies all alone in a bucolic landscape. I log a lot of solitary hours in some of these activities myself. If I worked at it, I could spend most of each day alone. I'm not saying all famers are introverts. I'm just saying the job might tend to provide lots of time for introspection. This is surely not the kind of life for people who compulsively plan parties or love committee meetings.

Many folks assume I'm extroverted because I'm often called on to speak in front of groups, entertain groups or step up in public forums. I majored in theatre and worked for years in radio, so people see me as a lot more outgoing than I am over the long haul. That training has paid off over and over, allowing me to function in an extrovert-dominated world, and in situations where I need to assert myself. But it's not my native propensity. And though my personality has moved more to the center in the last twenty years or so, I still definitely lean more toward the introversion side. I do need time by myself to regroup after steady periods of social interaction. In fact, I could spend days alone in my house, happy as a clam. In the winter, I sometimes pray to be snowed in so that I'll have an excuse not to leave the property for a long time. I go to the grocery store only out of sheer necessity. I would become a bona fide hermit, if my life did not regularly force me out of it.

So imagine my bemusement when I find myself inviting people out to the farm, as often as I can, hand over fist. *Come knit! Come crochet! Come spin! Come take a class! Come meet the animals! Come help me bathe the chickens!* What am I thinking? Do I really want all these people hanging out on the property? What if they won't go home? Where can I hide?

It seems to me there is another Hand at work here. I blame God. I believe God wants me to come out of my shell and love people. And He has an amazing sense of humor, asking me to do the job of a party planner. But He's right. People need love, and I have it to give. They need each other's love, and they need to get together in order to share that. So I've had to grow and adjust. I've had to open up and roll out the red carpet to my heart.

In committing to stretch a bit out of my comfortable, introverted nature, I had to risk some very scary possibilities. I had to let down my guard and risk vulnerability. As the "leader" of the LRB community, I felt like I needed to be strong, needless, and perfect. After all, people were depending on me to keep everything chugging along, and they needed a

dependable captain. How could I dare show my weaknesses? How could I ever not have the right answer?

Paired with my introversion is a strong need to control. But anything I controlled could only grow as large as my own effort and abilities. And I wanted the farm to grow. So I had to let go and let other people in. I had to delegate and share responsibilities with the folks who actually volunteered in different areas. When I delegate, I also must lower my unreasonable expectations. No one is going to do my jobs like I would do them, or as well as I. After all, I probably invented the job and have honed my skills over time, so how could anyone else do it to my specifications? Won't the world crumble if someone lets me down? Backed into the corner of wanting more from the farm, but not having the resources to pull it off alone, I was forced to ask

for help, to delegate, to supervise, to cheerlead my helpers and to hold my expectations lightly. This was not going to be easy.

All this stretching and opening up has changed me. I still need alone time, but I welcome together time, too. My friendships have grown stronger from the times of working shoulder to shoulder and successfully tackling tough challenges together. I am under no delusion that I have developed this farm all by myself. I am delighted to be just one of the many unique contributors.

No Cult of Personality

Some have suggested that it is something extraordinary in my personality that has made the farm and the Barnie community what it is. I must protest. I am not extraordinary. At least, not in the sense that many folks use the word. In fact, my shortcomings are legion. I am terribly disorganized, despite my pressing need for order. I procrastinate. I have great ideas but lack a lot of follow-through. I am a great starter, but a terrible finisher. (I hardly dare admit to all the knitting and crochet projects that lie around in bags, started but not even close to finished.)

These faults come from having lots of amazing ideas burst out of my brain—so many that I can't accomplish them before my passion cools for those ideas. I love to see how a thing might work—how a yarn knits up, for instance. But once I get a good idea of how that idea is going to work, I'm done with the project, even if it's not complete. Is it because there are not enough hours in the day? Heck no. If time-wasting were an Olympic sport, I'd be a medalist.

I'm sure I can't even see all the ways I'm not perfect, but here's what I do see: my introversion forces me to take breaks

from people more than I'd like. I can be very impatient. I have it in me to be snippy and sarcastic. I love a clean house, but have a wicked bad time keeping mine clean. Here's a terrible admission: some days, I secretly wish I lived in an apartment, with no four-footed mouths to feed. If I didn't make appointments, I'd never get clothes on or get out the door before noon. I'm actually pretty shy about meeting new people, and nervous when talking to strangers on the phone, especially when I need to be persuasive. Accounting makes me break out in hives. That's the short list.

I have succeeded here because I've been able to put my finger on all these flaws, drag them out into the light and address many of them by the airing. I've been able to recruit people to come along beside me and make up for the holes in my abilities. I have been remarkably blessed to find people who are skilled in areas where I have no skill, who are willing to hold my hand and take off a lot of the burden. Where my skills are mediocre and a job would take me days, I have friends who can whip it out quickly with very little expenditure of effort. Sometimes they just take the horrible duty and do it for me—oh, the luxury! Sometimes they come show me how to do it and hold my hand while I try. Sometimes they make an appointment with me to *come sit with me while I do it,* so that the stupid thing actually happens. For some reason, good company helps me to push away from the computer and focus on the items on the To Do list. It feels really immature to need this kind of help, and I'm not proud of it, but there you are. I'm a hot mess in many ways, and I couldn't do this farm thing alone for even one day. Gratitude swells up in my heart when I think about all the effort that has poured into this farm on my behalf. I can hardly contain it. Really smart, generous people have come to my aid, and the aid of the farm, and I am humbled and grateful.

So even as I'm sometimes tempted to beat myself up over these character flaws (as they seem to me), I see that my weaknesses have left a gap for others to come fill in. If I could do everything perfectly, I would have no need for others, and community would never develop. My places of lack have become places where others can accomplish. Where my light is dim, others can shine. Where I don't have a clue, others can be the heroes. I'm learning to accept my lapses because they leave room for others. Needing help has allowed me to give my friends a wonderful gift—the gift of *being needed*. Perfect and competent people need nothing, so the people around them have nothing to contribute. Perfect and competent people don't know how to give or take.

Here's what I think I do right: I say thank you. Knowing how much I owe my friends generates a mountain of happy gratitude. The very least these helpers should receive is repetitive verbal acknowledgement of my thanks. Sometimes we barter things of value for the exchange of work and time. Sometimes I'm able to pay a tiny wage. I try to always show respect for people. We may be different in fundamental ways, but I will always respect you as a fellow human being, especially if we share an interest in the farm and our crafts. I know my helpers' time is valuable and I try to respect that. I love to applaud people. I love to draw attention to their little acts of kindness on my behalf and on behalf of the farm. I love to applaud someone else's art. The products of our creative efforts mean the world to us, and I want to be sure that the Barn is a safe place to show them off among trusted friends. Everybody needs more "attaboys." Everybody.

I'm positive. I look for silver linings. I smile when I greet you. I'm happy to shake your hand or hug you if we're close friends. Some of this behavior is a factor of personality, and some of it has been learned. I watched what people did that

Giving Away the Farm

made me feel important and valuable, and I tried to copy them. I know that how you feel after you've been to the farm will determine whether or not you'll want to come back and be a part of our community.

I knew the Barn community had turned a corner when I had occasion to be away on one of our usual Saturdays. It was time to let the group take care of itself. I left some instruction and tidied the place up a bit. I asked a couple of people to serve as hosts in case we had any visitors. I went off on my trip with full faith that the community would manage without me. And they did. They sent me texts with photos to show me what was happening during several points in the day, which pleased me no end; they didn't want me to be left out. And I welled up with pride and gratitude that our little bunch of friends had effortlessly taken up the torch in my absence. Coffee was perked, eggs were sold, newcomers were greeted warmly, projects were shared, and the day definitely went into the win column. I hate to miss LRB days, but if I have to, I know that everything will run smoothly.

This group of friends may or may not be your perfect fit, but you won't leave because the Shepherdess wasn't nice to you. That's my goal. That is my fervent desire. I also hope that if I can consistently model this attitude, it will rub off on every Barnie that greets a newcomer, and the LRB will become a place that just nearly glows with love.

CHAPTER TEN

Shepherd Interrupted

The life of the Little Red Barn has exceeded our dreams. I always find intriguing new friends, encouragement, caring and listening ears, the belonging and connection our neighborhoods often lack, inspiration as a creative, and deep rewarding Joy that I am loved and needed here.

—Laurie M., Barnie

giving Away the farm

As the Steward of the farm and as the Shepherd of the Little Red Barn, it fell to me to model Graceful Neediness. And so I did. Fall, that is, and not gracefully. But with my spectacular slip off the front porch, all the Barnies' kindness, compassion, generosity and commitment poured out in a beautiful community response to my very dramatic need.

On December 14, 2012, I was busily about routine farm life.

My friend Taya and I had been working together to clean the chicken coop. She had driven with her very young son from her home about 45 minutes away to help me with this really unpleasant job. It was my only real agenda item for the day, so I was looking forward to a quiet afternoon. We laughed and shared and had a pretty nice day if you overlooked the chicken poo, dust and wood shavings covering our boots and jeans. The weather was fairly mild for December, and a light mist began to roll in as we were putting away our tools. This signaled to Taya to pack up her stuff and head for home, and lunch. A few minutes later, one of our Barnies' husbands pulled into the driveway to purchase a CSA share for her Christmas present. We chatted for a few minutes and then he left to finish more holiday shopping. As I walked back into the house, I noticed the alpacas in the north pasture doing something particularly silly or noteworthy, so I ran inside and grabbed my phone to use the camera to capture the moment. I stepped back out onto the damp porch in the drizzling rain, and felt my Croc-shod right foot slide out from under me.

The next two seconds seemed to go by in slow motion. I remember frantically trying to catch myself, knowing the hard flagstone walk was not going to be a comfortable place to land. I heard Judah and Tella barking, and realized that they were reacting to a sound—the sound of me letting out an involuntary wail. I did a terrible job of breaking my fall with my left hand and hip. I was down. My ankle hurt badly, and I was instantly angry at myself for twisting that ankle yet again. Every time I had done so previously, it meant hours on the couch with my foot in the air wrapped with a frozen ice pack. I really didn't have time for that kind of nonsense this close to Christmas.

I stayed still for awhile, trying to breathe through the pain, and waited for it to subside enough that I could stand up and get back into the house. But it didn't subside. Not only could I not get to my feet, I couldn't even move from my prone position on my hip and elbow. "Dang," I thought. "This is bad." I started brainstorming some options. Here's where I discovered the biggest miracle of the day: my phone was still in my hand. In the fall, I had not lost my grip on it, so I actually had a way to communicate. I called my friend who lives about a mile away to see if she could help me up off the sidewalk and then pick up my daughter, Emma, from school at the appropriate time. She didn't answer. I left my pitiful message. ("I've fallen and I can't get up." Really.) And then I thought some more. Ultimately, I called my husband, Ted, who called 911 on my behalf. So embarrassing.

Parker's Finest arrived first. When the police chief got out of his car, I had to raise myself up on my elbow and wave for him to see me behind the salvia bushes. Shortly after, two more police cars arrived on the scene, and down the drive came two officers I know by first name. *I love living in a small town.* They asked me a lot of questions *(had I lost consciousness? did I hit my head?)* but I assured them my main injuries were to my ankle and my pride. Soon, a paramedic arrived and the group lifted and supported me as I hopped up to the porch and into the wooden rocker. They elevated my foot and immobilized it with an inflatable cast. One of the guys stepped into the house and pulled a knitted

Giving Away the Farm

lap blanket off of a chair to put around me. I had almost forgotten the drizzle and the 50-degree temperatures that had blown in as I lay on the flagstones. Crazy thoughts went through my head while we waited for the ambulance: *Thank*

God I don't have to go to the bathroom! Who's going to feed the animals tonight? What does an ambulance ride cost? What's going to become of Christmas? (Control issues much?)

In the midst of my mental gymnastics, I somehow decided I needed to use my phone to take all the officers' picture. After all, I've never had that many uniforms in my front yard, and these guys were really pouring on the Dashing Rescuer routine. They hemmed and hawed, but ultimately submitted to having their photo taken. As long as I kept my foot still, we could all laugh and joke. I was working hard to cover up my embarrassment and the fears I could not yet articulate about how all my work was going to get done that evening and beyond. Finally, I surrendered to the reality that I couldn't do a dang thing about my situation. The guys were very good to me, and I relaxed into their care.

Another EMT and I rode to the hospital in the back of a cold, sterile, bouncy ambulance. I don't know what I expected an ambulance ride to be like, but it turned out to be singularly disappointing. At the hospital, the crew rolled me in through the emergency room entrance, and I swear the medical team that met me was disappointed I wasn't a gunshot victim. They were visibly let down. I wasn't even bleeding! I maintained my characteristic jolly demeanor, laughing off the fact that I had just taken an ambulance into town because of a twisted ankle. What an overblown annoyance.

The first stop was radiology. I impressed the technician with my unusually high pain tolerance as she moved my foot and leg this way and that to get the best pictures. I did get some good medicine for pain, which I'm sure made a difference, but from the looks on everyone's faces, I began to deduce that this was worse than just a twisted ankle. Staring up at the ring of doctors and nurses around me, I got a sense of what my occasional sick or injured animals go through.

Giving Away the Farm

And I decided I prefer being on the care-giving side of the table, not the care-receiving side. I don't do "helpless" very well at all.

The orthopedic surgeon confirmed a fear I had not let myself even consider: breaks. Three places. I'd need surgery where they would re-attach some ligaments and drill screws through some bones that had chipped off. They would just leave another fracture higher up my leg to heal on its own. Apparently, I had worked very hard to keep myself from falling, and had done a bunch of damage in the process. I'd be in the hospital five days or so, and then would not be able to bear any weight on my right leg for more than six weeks. I'd need in-home occupational and physical therapy. Later, I would be able to bear a little weight, but wouldn't be well enough to do chores again until the spring. Wait, what? *Spring?* Who was going to do all my work? How would my livestock survive? *What about the farm?*

My husband turned to my daughter and said, "You're getting your license over Christmas break." Yes, sir, everybody was going to have to step up.

For a couple of weeks, Ted bore the lion's share of the burden of animal chores. He enlisted Emma whenever he could, but it was mostly his duty. I did my best to write out the daily jobs and explain how I was used to carrying them out. Christmas preparations pretty much screeched to a halt. Fortunately, I had gotten decorations up before the accident, but plenty of gift shopping simply went undone that year.

My church has, over the years, developed a pretty slick way of getting temporary help to families who need it, particularly in the case of new babies, illness, injury or a death in the family. Before I knew it, my friend Kris had organized the online calendar with places for people to sign up to bring casseroles or to handle animal responsibilities. Ted trained a small crew of Barnies who relieved him of a good portion of the work. Immobilized in bed, I could look out the window and watch my friends feeding the sheep, alpacas and dogs on the north end of the property. My eyes welled with tears every time. I watched sweet Jerry and Eunice, Kate, Gail, Bryan and Elizabeth and others carry on in my place so that everything ran as smoothly as possible.

From my blog, December 28, 2012, after two weeks of living as a handicapped person:

I'm still trying to figure out how to live [handicapped.]

This is the life that lots of folks have to endure temporarily or permanently. I've never known before just what that's like. I'm gaining a big appreciation for handicap access features—hand bars in bathrooms, curb cuts in sidewalks, wide spaces between parked cars, etc. And every day I realize how much I take normal ambulation for granted.

But my hardest lesson has been learning to receive—having other folks do my chores, help me dress, wait on my slowness, wrangle my equipment ... it's so hard to be the non-productive cog in the wheel. This appears to be how I judge my value—by my productivity, and I know that's not right.

By God's grace, I've also been learning other lessons over the past several years—that we create something of value when we love another person ... when we create a space for

people to come together, and care for each other, and express their gifts. Our community here at the farm has grown so much in the past several months. More and more people have been drawn to this kind of loving, creative community for what they can give to it, and for what they can get out of it. This sounds kind of vague and misty unless you've experienced it firsthand.

And I'm enjoying the benefits of our community in a very, very concrete way. Thank you, thank you, thank you to all of our sweet friends who have helped Ted with chores, brought food, called to cheer me up, and more. I love and treasure you all more than I can express. Thank you for helping me on this journey of waiting, and receiving.

Finally, after what seemed like an eternity, I was able to get back to my scheduled duties. Though I was briefly tempted to feel a) guilty for making a lot of work for people, b) guilty for secretly enjoying an enforced "vacation" from chores, c) distrusting that anyone would be able to do my chores well enough, and d) offended that they actually had done my chores adequately, I opted for e) grateful—overwhelmingly grateful that I had talented, loving, committed friends who were, for a specific period of time, able to take up the mantle and care well for the farm. All the critters had fared just fine, and the team kept the farm clicking along on the busy spring calendar of birthing lambs, shearing the flocks and processing fiber—mostly without me.

All that time I had spent in bed was put to good use, as I was able to plan strategically for the farm's future in a way I had never had the luxury to do before. When one is stuck on one's backside, the smartest thing to do is to apply the brain and the imagination. I also got a lot of knitting done—a bonus I did not take for granted.

The accident inspired me to cut back a bit on the number of animals I had to care for. As much as I loved them all, I didn't want anyone to have such a big job if, God forbid, any other accident or even a vacation away kept me from my chores. This situation helped me redefine for myself how many critters is enough for our farm. We really only need enough fleece and fuzzy faces to make the place a joy to visit.

A couple of years ago, I toyed with tweaking the whole idea of CSA, Community Supported Agriculture. Instead, I wanted us to be known as an ASC, an Agriculturally Supported Community. I exploited the play on words of being knit or woven together with wool and fiber. But then I pushed the analogy nearly to the limit: I thought about wool felt and how it's made—it's the rough parts of fibers, agitated together in hot water and soap, which then interlock together into a seamless fabric. Kind of like us. Locked together in friendship through lots of interaction of ideas, trials, victories, hilarity and creativity. And this is so much greater than any dream I had ever dared to imagine.

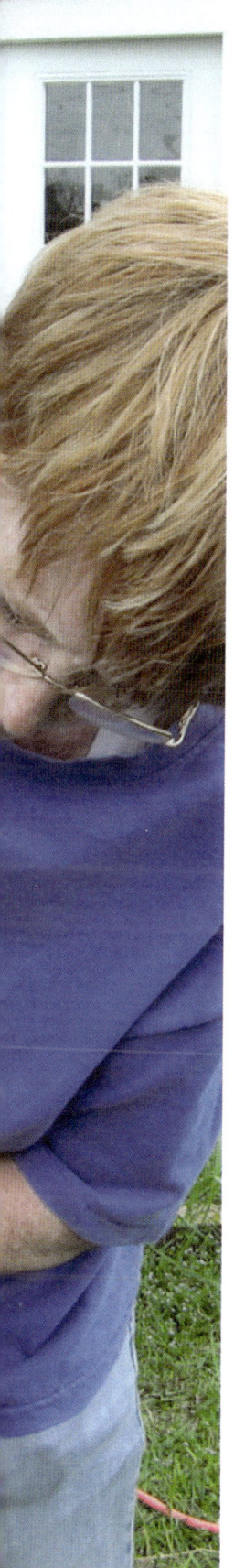

CHAPTER ELEVEN

Generosity and Learning to Receive

I love the oohs and aahs over finished projects, even when I was less than thrilled with my output. I love the way the conversation easily moves from family to religion to politics to TV to pop culture as well as all the fiber-related topics that brought us together in the first place. I love that we never seem to sink to the level of gossip and malicious conversations about people. I love when I find out that each person I meet at the barn has at least one more thing in common with me than the fiber arts.

—RITA, BARNIE

Giving Away the Farm

One of the toughest things for consumption-driven Americans to turn loose of is our Stuff. We spend tons of money on stuff, and then we store it, clean it, maintain it, worship it, move it around, polish it, defend it, waste it, damage it, lose it, insure it and hoard it. Real freedom comes from letting go of stuff—even sentimental stuff, or stuff that still has value, or stuff that somebody gave us, or stuff for which we paid a lot of money. I've heard it said that the best antidote to greed is generosity. So if you feel like your possessions possess you, it's time to give some of them away.

People who give are secure. They don't need stuff to feel good about themselves.

People who give have eternal values. They know this stuff will all pass away, and we can't take it with us.

People who give experience the joy of seeing it used by people who actually need it.

People who give free up space (physical and mental) to actually give them peace, so that their attention can focus on consequential things.

The fruit of all the hospitality, acceptance, respect and compassion practiced here at the farm has matured into constant acts of generosity. When we feel full and satisfied in our personal relationships, we often find ourselves inspired to give to others. So in the loving, caring culture of the LRB where friendships thrive and resources are shared, the obvious outflow is giving.

The farm itself was the first recipient of this giving—from very early in our history, Barnies and Shareholders have given hours and hours of their time, helping with unglamorous cleaning and maintenance jobs, animal doctoring, fiber processing, and various administrative tasks. They have donated money and supplies to provide the things we desperately need to run the farm, not the least of which were

Generosity and Learning to Receive

almost all of our livestock guardian dogs, lots of equipment, some fencing, and more.

At first, all this receiving was pretty tough for me. As a for-profit business, I felt like I shouldn't take financial gifts or high-value physical gifts. I couldn't give anybody tax deduction forms or anything else in return but my gratitude. I wanted to make sure no one felt used, and I, truth be told, didn't like feeling beholden. (How selfish is that?) But my attitude softened when I mentally put the shoe on the other foot. What if I felt part of a community, a family—whether for profit or not—and I wanted to contribute my gifts or abilities? Would I feel valued as a contributor to the greater good? Would I feel needed and appreciated for what I could chip in? The answer, of course, is yes. My fears and discomfort as the receiver needed to be tossed out like the rubbish they were. They stemmed solely from my pride. Was I too good to receive? The farm, and all it represented in the lives of the Barnies, was payment enough (they tell me) for the gifts they offered. I needed to get over it.

I want to honor and value each of our friends for who they are and for what they love to share, so I quit saying, "Oh no, that's not necessary," and now say instead, "Why thank you, that would be a big help and a wonderful blessing." The Barnie receives by giving, and I give by receiving. Cool, how that works, isn't it?

Farming Takes a Village

In the early days of the farm I tried to do my own sheep shearing. I'm a can-do kind of gal, and if I could learn to spin, surely I could learn to shear. I started with hand blades, but got so nervous and frustrated after destroying a couple of fleeces that I asked for any help I could get. Moral support

makes all the difference. Next, I invested in electric shears and a couple of different shearing stands, hoping these might make the job more manageable for a newbie. And in the spring of 2009, shareholders Chris and Kate, along with my husband Ted, came along for the adventure. Chris is a scientist and loves the vet stuff, so she volunteered to give the yearly vaccinations, and Kate is a spinner and animal gal, so she was very helpful in wielding the shearers and handling the sheep. Ted is a tall guy with strong hands who operated the hoof trimmers with relative ease. My job was to have all the supplies ready, supervise and jump in anytime I could. Between the four of us, you'd think we could have managed four sheep.

At 9 A.M., my friends arrived, and after some last-minute organizing, we started our shearing. Ted, Chris and I started on Tommy the Jacob sheep because we figured he'd be one of the easiest fellows to tackle; he was the oldest sheep we owned, and had been through this experience every year. My makeshift shearing stand was actually a wooden, homemade goat milking stand that I got on Craig's List as an replacement for the metal stand I got on eBay with a faulty head piece. (It's always something.) Tommy's shearing went so well that I forgot to get my camera out to document it. "Wow," I thought, "this year is going to be a piece of cake."

Next up on the stand was Israel, the larger of the two Jacobs. As we got his wool off, we could see that he had been eating just fine, thank you very much, and all those plaintive cries for more food had been just pure greediness. But even though he was nicely fleshed out, he looked so small without his fleece! Using the electric clippers really improved the finished look of those two sheep. Much less of the "Weed Whacker Special" style they had gotten the previous two years. As usual, we collected a large plastic bag full of Israel's

Generosity and Learning to Receive

charcoal gray, soft wool. After each sheep got his haircut, he also got a tetanus vaccination from Chris, and a pedicure from Ted. Full spa treatment, so to speak.

Shareholder Kate arrived and joined in the fun. It was time to drag out big, fluffy Shadrach, the young Suffolk wether. Shad had never been sheared before, so we knew he'd have to learn how to deal with this obnoxious new activity. Early into his shearing, we noticed that the blade in the electric clippers was beginning to dull. The job got much harder, and it was clear that the day would be longer than I had anticipated. My arm and shoulder were getting really tired from pushing that dull blade into the wool over and over. Ted, Chris and Kate helped out by using the hand shears when they could, to attack the fleece from different angles.

Then it happened.

I realized as I was grinding away on Shad's back end, that I had inadvertently taken a pretty good-sized piece of skin off with the last pass of the clippers. OMG. Now, I have nicked a sheep in the past before, but this was a big hole. It wasn't bleeding badly, but I couldn't decide the best course of action. The thought had occurred to me earlier that I might like to have a tube of SuperGlue at the ready, but I had totally spaced it out. Chris reminded me (my head was spinning and I was kind of nauseous) that I had TriCare ointment handy, which might be a good idea—it works as a wound cleaner, antibiotic and pain reliever. I took her advice, but as time went on, I really didn't like the look of that hole. Chris, vigilant on vet tech duty, suggested that maybe I could sew it up. "Oh sure," I said. "Why not." So I dashed into the Red Barn for a curved needle and black sewing thread, thinking that black thread would be easier to find and remove later. I put a few stitches in to close up the hole, praying and trembling. I've never done that

before. Wow. I did it. Wow. I'm not sure I would have had the courage to do that without the crew cheering me on. I still felt a little sick, but I was so much happier with that gaping hole closed. Thankfully, Shadrach hardly blinked as the whole operation was going on. That helped a lot, too.

But the gash took a toll on my confidence, and I found myself really having to *make* myself continue with the shearing. All of a sudden I was afraid. But again, I was surrounded by great friends who encouraged me, and I pushed into my doubt. It wasn't the last nick I delivered to a sheep that day, but I kept plowing. The job had to get done. Kate really helped a lot by taking turns with the clippers herself. Chris helped with the hand shears, and Ted was right there in my corner.

Once we got Shadrach back into the pen, I decided it was time for a break. Kate and Chris and I retired to the Red Barn for the traditional quiche and coffee. It felt so good to sit down. Once we got our energy back, we jumped back up to finish Sheep Number Four—Zacchaeus, the Babydoll Southdown ram.

Now, Zach (the Tiny Tank) had never been sheared before either. We got him out of the pen easily enough, but once out in the yard, he decided that "playing dead" was his coping method of choice. He lay down on the grass and closed his eyes. We had no choice but to get the green garden wagon, hoist him up into it, and head for the shearing stand. However, once in the wagon, he changed his tack and tried to jump out. Ultimately, we sat him on his butt, and I held him upright while Kate pulled the wagon across the yard. (A YouTube video of this episode would have made us all rich.) Now, Babydoll Southdown sheep—a breed that corners the market on cute—have wool all over every inch of their bodies except their eyeballs and their hooves. And I mean *everywhere*. Zach's dense, sponge-like fleece would prove to be a completely new kind of challenge. Once we got him near the shearing stand, we decided to take the clippers to his tummy and inner thighs right there in the wagon, since we had him in the perfect position. He wiggled a little but mostly he laid his head on my shoulder, closed his eyes and prayed a little sheepy prayer that all this would just go away.

We moved him (no small feat) to the shearing stand to work on the rest of his body, and at that point he decided he'd just lie down again, tucking his four legs under his roly-poly body. All of us intrepid shearing ladies were getting tired by now, and it was creeping on into the afternoon. Kate and I took the shearers to his back, sides, and as much of his neck as we could reach, but it was clear that the electric clippers had gone beyond the point where they were effective. As we took the wool off his back the heat that radiated off him could have poached an egg. He must have been so relieved to be rid of it, in spite of his ordeal. We picked at him for a little bit with the hand shears and then we ran up the white flag. We pulled and pushed him back to his pen and turned him loose. Frankly, he looked atrocious, as though he'd been shorn by a blind landscaper with a hedge trimmer. Most of our hard work lay in clumps of short wooly bits all over the yard rather than in the fleece bags, but we were done. We were finished, in every sense of the word.

That afternoon, I accomplished more than my first sheep surgery. I learned that sheep are not nearly as delicate or fragile as I had thought. I learned that I can be pretty tough when I need to accomplish a scary job. I learned how big, scary jobs turn into treasured memories when shared with friends. I learned not to hold up independence as the highest quality for which to be known, but interdependence. I learned to share the work and the wonder with others.

Working the Wool Together

 This is a fiber farm, and so our central product comes off the backs of our sheep and alpacas. Once the fleeces are sheared, they must be processed, every year. As I've mentioned, I love the journey of the wool from sheep to shawl, and in small doses, it can be quite therapeutic. But when you have 20 to 30 fleeces to get squeaky clean, and shareholders who are patiently waiting for their booty, the sheer volume can overwhelm even the hardest-core fiberista. Some of these generous shareholders who have already paid good money for the fiber quite often help out by rolling up their sleeves to get the product marched through the labor-intensive cleaning process so that they can finally enjoy their reward.

 Wool and alpaca fibers are very different from each other, and require different processing steps. Both kinds of fleece begin their journeys by being spread out on a "skirting table" (a mesh-covered flat surface), so that the soft, desirable fiber can be picked over and separated from the coarse, undesirable fiber and other contaminants. Any sticks and stickers, dung or other nasty bits are pulled out and discarded. This "skirting" takes practice and some time. It's a job for eyes and fingers, trained by repetition. Toss out too much fiber and you decrease the amount of your finished product. Toss too little out, and you degrade your finished product with coarse or short fibers that coil up and pill or give the yarn a scratchy overall feel. Over the years, I've decided to err on the side of pulling out every questionable bit of fiber in order to be left with only the best. This helps us assure shareholders a quality product. We always have plenty. The animals will grow more next year.

 At this point, the two fiber processes diverge.

Sheep wool is full of sticky lanolin, which adheres the pasture dirt and vegetable matter to the wool fibers like glue. This lanolin has a waxy base which must be melted out of the fiber with super-hot water and grease-cutting soap. Each fleece is soaked in this hot water and soap mixture a couple of times and then again several times in clear water to rinse it. Once the wool is really clean, it is air-dried. (I've found this site to be a great resource if you want more information about wool preparation or spinning: http://joyofhandspinning.com/how-to-wash-your-fleece/.)

Alpaca fiber has no lanolin to hold on to the dirt and hay, but unlike sheep, alpacas love a good roll, which buries the debris deep into the fleece. To dislodge this contamination, we "tumble" the fleece in a rotating drum covered in mesh that bounces the fiber around. (By now, you'll probably not be surprised to learn that this tumbler was gifted to the farm by a generous Barnie.) I combine the tumbling action with blasting from a leaf blower, and clouds of dust, dirt and short fibers fly out of the fleece as it rotates. Every bit of dirt that gets blown out is dirt that I don't have to wash out, and this speeds the process immensely. Then, the alpaca fiber is washed much like the wool, taking care not to agitate it, which might cause felting. The fiber is then air-dried. I like to use a four-level sweater dryer with a gentle, cool fan to speed the drying process. (YouTube has lots of informative videos of the fiber process. I found this one on tumbling and blowing llama fleeces that shows it pretty well. No affiliation: http://youtu.be/H0kda0W49Ao)

Dry wool and alpaca fleeces are then put into plastic bags with all the air vacuumed out. These compact "pancakes" are stacked in sturdy boxes and mailed to our processor. My friends Lynn and Jim Snell at Spinderella's Fiber Mill (www.spinderellas.com) in Salt Lake City have cared for our fleeces

for years, and the results have always been spectacular. The Snells card the fiber on their vintage, industrial-size carding machine, and it comes back in a rope-like preparation called "roving," ready to spin. We divide the roving up into the appropriate share size and distribute it to the shareholders.

This is the ideal process. Several times we've experienced bumps in the road getting the fiber clean and off to Salt Lake, and in every case, Barnies have come alongside to help me get the big job done. They take home some hands-on learning and experience, and I get some welcome help with the labor and the responsibility. They do it because it's fun, and because they are generous.

Generous with Each Other

As the friendships grow deeper, generosity overflows between the Barnies, as you might expect. Yarn is the currency of knitters and crocheters, so regular de-stashing goes on in the LRB. Sometimes folks go home with new yarn for themselves. Sometimes they just end up swapping old yarn for "new-to-them" yarn from someone else's stash. Many times, the new acquisitions become finished items for charity. All is freely given and received with love and joy.

The Barnies have struck up such friendships at the LRB that they have spilled over into "real life." The ladies get together for lunch or drinks or to cat-sit for each other when needed. Dye days or batt-making days spring up at individual Barnies' homes, attended by other Barnies. Acts of kindness don't have to be expensive—they can even be symbolic tokens of friendship. We learn things about each other that inspire us to give: Chiyo loves pink, Denise loves rubber ducks, Rita is partial to Winnie the Pooh—knowing these personal details about each other keeps us connected.

When I started hearing these stories of extra-barn time get-togethers, my heart welled up with joy. This is the point, to my mind. The point is to take what we have together in the LRB out to the world and share it. I'm very gratified that LRB friendships have the integrity to extend beyond our walls and to enrich the Barnies' lives in a more comprehensive way.

And when one of our Barnies succeeds, especially in the yarn world, the community responds with whoops and cheers. Our friend Denise is frequently a competitor at the State Fair of Texas with her crochet projects. Recently, one of her amazing crocheted bags was featured in a beautiful art book.[5] Our friend Greta is a recognized and celebrated visual artist with pieces in distant galleries. Our friend Laurie is a knitwear designer who self-publishes and who also has designs published by large yarn companies. Peggy is a writer who regularly publishes feature articles we can enjoy. Many of our friends sell their wares through their own websites or Etsy shops. Some Barnies are encouraged by the group to reach big goals that stretch them beyond what they would normally attempt, whether it's learning a new stitch or starting a new business. All these announcements of achievement are met with applause and delight—just like when we each hold up our latest stitched creation.

One of our most notable instances of Barnies putting their heads together to make something magical was the teaming up of Laurie's pattern and Ashley's gorgeous hand-dyed yarns on the occasion of the second annual DFW Yarn Crawl. Laurie's cowgirl-inspired "Barbed Wire" cowl teamed perfectly with Ashley's "Hard Day's Work" color way to form the perfect Yarn Crawl project. Ashley could hardly keep her yarn stocked in the shops during the Crawl, her colors were

so appealing. And in yet another act of generosity, Laurie has allowed me to share her pattern with you. Pages 103-105.

Generous to the World Beyond Our Circle

Many yarn crafters give to charity. They donate both yarn and finished items to causes like NICUs in local hospitals for the tiny premature babies. They make prayer shawls for those going through rough times. They craft hats and dolls for cancer patients, young and old. In these ways, the Barnies are just like most yarn crafters you know. But they go further.

We have an ongoing call for yarn that periodically gets transported to a crocheter in the Houston area who makes afghans for poverty-stricken people in Guatemala. Photos come back to us, and we can see our yarns that have been made into beautiful blankets, and the faces of the adults and children who have received them with thanks. It's a gift back to us to be able to see our tangible love in action. We have collected scarves to give out in cold weather to the homeless on our own streets. Several years ago, the Barnies collected enough cash to donate four flocks of sheep internationally to needy people through World Vision.

Over the past couple of years, several Barnies, including Michelle, Anela, Victoria, Nancy and Kris, have helped out with Farm Camp, allowing us to share the world of the farm with kids between seven and ten years old. These same friends have helped me lead school tours around the farm for scout troops, home school co-ops and families. These Barnies are eager to share what we have here at the farm with young minds—the minds that will steer the world after we're gone.

Barbed Wire Cowl
By Laurie Beardsley

Barbed wire? Around your neck? Seriously? If it's this lovely cowl, you can. Scrunch the small version up near your neck, or wrap the longer version multiple times for extra warmth. This cowl is knit in the round. It features a traveling stitch motif on a purled background.

Sizes
Small: 24" (61 cm) circumference and 5.75" (14.6 cm) width
Large: 48" (122 cm) circumference and 5.75" (14.6 cm) width

Materials
- DK-weight yarn, merino or merino blend. Shown in Dye2Spin 100% Merino DK in "Hard Day's Work" colorway.
    - 140-175 yards (130-160 m) for size Small
    - 275-375 yards (250-345 m) for size Large
- US 5 (3.75 mm) circular needle, at least 16" (40 cm) long (36" (90 cm) long recommended for size Large)
- Cable needle (optional)
- Tapestry needle
- Stitch marker
- Blocking supplies (optional)

Gauge
23 stitches and 30 rounds = 4" (10 cm) in Barbed Wire pattern

Abbreviations
k = knit
p = purl
ktbl = knit through back loop
st(s) = stitch(es)
[] = repeat the instructions the specified number of times
LT tbl = left twist through back loop (slip stitch to cable needle and hold in front, ktbl next st, then ktbl st from cable needle)
LPT tbl = left purl twist through back loop (slip stitch to cable needle and hold in front, p next st, then ktbl st from cable needle)

Barbed Wire Pattern

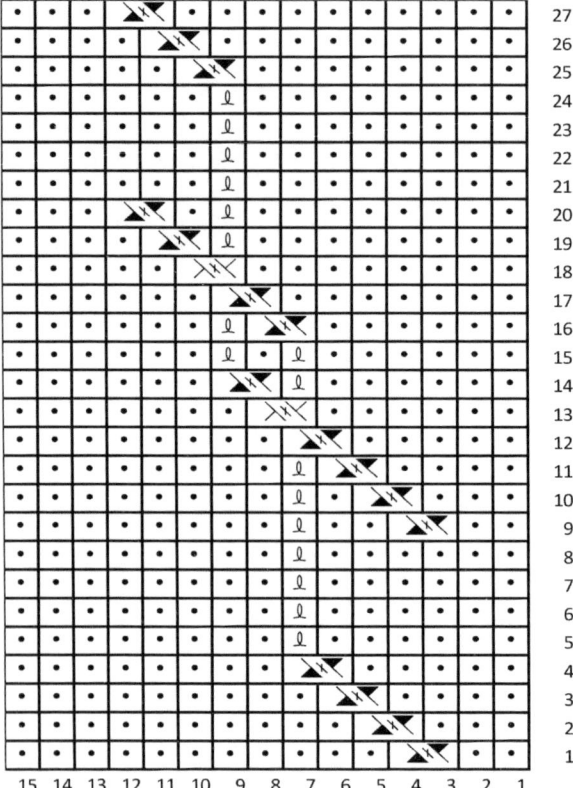

Row 1: P2, LPT tbl, p 11.

Row 2: P3, LPT tbl, p10.

Row 3: P4, LPT tbl, p9.

Row 4: P5, LPT tbl, p8.

Rows 5-8: P6, ktbl, p8.

Row 9: P2, LPT tbl, p2, ktbl, p8.

Row 10: P3, LPT tbl, p1, ktbl, p8.

Row 11: P4, LPT tbl, ktbl, p8.

Row 12: P5, LPT tbl, p8.

Row 13: P6, LT tbl, p7.

Row 14: P6, ktbl, LPT tbl, p6.

Row 15: P6, ktbl, p1, ktbl, p6.

Row 16: P6, LPT tbl, ktbl, p6.

Row 17: P7, LPT tbl, p6.

Row 18: P8, LT tbl, p5.

Row 19: P8, ktbl, LPT tbl, p4.

Row 20: P8, ktbl, p1, LPT tbl, p3.

Rows 21-24: P8, ktbl, p6.

Row 25: P8, LPT tbl, p5.

Row 26: P9, LPT tbl, p4.

Row 27: P10, LPT tbl, p3.

Shown in Dye2Spin 100% Merino DK in "Hard Day's Work" colorway

Cowl Pattern

The pattern is written for size Small, with changes for size Large in parentheses. If desired, the circumference of the cowl can be changed by approximately 2.6" (6.6 cm) for each repeat of the 15-stitch Barbed Wire pattern that is added or subtracted from the cowl.

Cast on 135 (270) sts. Join for working in the round, being careful not to twist, and place marker for the beginning of the round.

Bottom Edging:
    Round 1: K1 tbl, [p1, LPT tbl] to 2 sts before marker, p2.
    Round 2: LPT tbl, [p1, LPT tbl] to 1 st before marker, p1.
    Round 3: [P1, LPT tbl] to marker.
    Rounds 4-6: As Rounds 1-3.

Body: Work all rows of the Barbed Wire Pattern, repeating each section 9 (18) times.

Top edging: Work as Bottom Edging.

Bind off in pattern. Weave in ends. Block lightly if desired.

Resources

1. Thank you to Ashley Wolfe of Dye2Spin (http://www.etsy.com/shop/Dye2Spin) for designing this special colorway for the 2013 DFW Yarn Crawl.
2. A good video tutorial on twisted stitches, and techniques for knitting them is here http://www.knittingdaily.com/blogs/daily/archive/2010/10/20/knitting-the-twisted-stitch.aspx

Copyright © 2013 by Laurie Beardsley. For personal use only. Please do not email, copy or reproduce this pattern. Please send comments and questions to LaurieBeaKnitting@verizon.net, or visit the LaurieBea Knitting Circle group on Ravelry.

CHAPTER TWELVE

Spreading the Community Beyond the LRB

Going out to the farm is the highlight of my weekends. I come home feeling rejuvenated and restored and just plain happy.

—LISA, BARNIE

Giving Away the Farm

The one and only regret I have about the Little Red Barn existing as a physical location in time and space, is that I cannot bring everyone here who would like to be a part of this group. We've had to apply some creativity to accommodate even all the local people who want to participate.

Not long after our Barnie get-togethers became a regular occurrence and our crowds began to grow to nearly full capacity, the ladies suggested that I find a way to enlarge the Little Red Barn to make room for even more people. Unfortunately, knocking down walls or adding a second floor just weren't options. As I've mentioned, our flood plain location severely limits our ability to build or expand our footprint at all. But I could make the *month* larger, so we began meeting two Saturdays a month instead of just one. Some folks picked their favorite Saturday of the two to attend, and some opted to come both weeks, and somehow, we've always had room for everyone who stops by.

I'm also doing my best to make the "LRB experience" something that people can access in other ways. Telling the story of the farm in this book is one way. I want to be as transparent as possible so that you can replicate the community you want in your own context. I want you to know, and see from my own example, that you don't have to be perfect or have a perfect situation to make this work. I want to encourage you to use the skills you have and the circumstances you're in right now as the basis to build your own community.

The internet is a fantastic tool. The Barnies have groups on Ravelry (http://www.ravelry.com/groups/jacobs-reward-fiber-farm), on Facebook (https://www.facebook.com/JacobsRewardFarm), and on Meetup (http://www.meetup.com/JacobsRewardFarm/). Through these avenues we can visit and converse over any amount of distance. We can advertise our existence and our openness to others through these sites. In fact, the LRB has a "sister" crafting group in the UK, thanks to the wonders of social media and email. I met our friend Caroline on Facebook, and we swap stories and experiences through posts and photos. Attempts to replicate the "LRB experience" in remote locations will not be exactly the same thing as being here, but perhaps that's actually better—each community will develop out of the personalities and goals of its own members.

You Can Do This, Too

Now, community doesn't just happen. It starts with one passionate person. Somebody has to stand up and say, "I'll start." Are you that person? Would you like to create your own "LRB family?" Your own community must reflect your own values, so it must grow up in your own place. Your own environment, whatever that looks like, will shade and color your group, and will enhance its unique sense of "home." Make it your own, to serve the people who are drawn to you. Since true community happens between people, an actual "farm" context is optional—the human element is the one non-negotiable feature. Are you a yarn artist? Great! You can take advantage of yarn's natural magnetic qualities to begin to draw a community together. Do you love books? Start a group around your love of reading. Gardening, pets, travel—all these have the potential as the glue to hold friends together. Be prepared to share your knowledge and your heart.

Is the small fiber operation the part of the farm that is appealing to you? Find a mentor. As we live on the land and learn from our experiences, knowledge collects as an asset to be shared with folks hungry to learn. No one wants to reinvent the wheel, especially a wheel that folks have been

using since the dawn of civilization—and I don't blame them. Farms have existed forever, and farmers can pass down what they know. Google is my friend; it has put an enormous amount of information at my fingertips along this journey. But I prefer to mine the firsthand experiences of working farmers. Then I like to try things out with a mentor and my own hands before I set off to work alone.

For that reason, I have offered classes in sheep and chicken keeping, knitting, spinning, weaving and crochet. A significant number of visitors come to see how we live the hobby farm lifestyle as they design their own futures on their own land. Not that we know everything there is to know, but we have learned some good lessons, and we can help folks avoid unnecessary mistakes.

We will be developing more opportunities to share our experience here because people need it; hundreds of people, young mostly, are returning to a more basic, rural life. My friends, the Horaks, helped me learn about sheep keeping by firsthand observation and experience. In a similar way, Ted's training when I broke my ankle gave several aspiring shepherds the opportunity to try their hand at the routine of daily chores. (As a side benefit, I now have the luxury of calling on a couple of well-trained friends to hold down the fort if I need to be away from the farm for a day or two. That's what you call a gift that gives back!) I'm excited to be a mentor to new young farmers coming up in the field (no pun intended).

Stretching the Little Red Barn's Footprint

Since we can only accommodate a finite number of people in the LRB, I set to work devising ways to extend the boundaries of the farm and our community. I've developed events where area artisans can come and hobnob and find an audience of consumers looking for their art. I can share with these artisans our assets of a lovely location and a network of fiber- and craft-lovers who are happy to support great local artists. One of the most valuable things I can offer our friends in the yarn arts is a way for them to showcase and profit from their talent. We have developed several avenues to achieve these goals, and the ideas just keep coming. One thing always leads to another.

"Share the Harvest" Fall Gift Market—Connecting Craftspeople and Consumers

About five years ago, I had settled into a nice routine here at the farm. We had hunkered down and survived another blistering hot Texas summer, and that energy-sucking weather had begun to let up a bit. I felt my strength and enthusiasm returning. I really wanted to capitalize on the cool, fresh breezes that were just beginning to blow across the pasture, encouraging me to spend more time outside. The time seemed perfect for a party on the lawn. I set up some tents, invited Alta, my chef friend, to cook us a huge pot of chili, and invited a bunch of fiber friends to bring a dish to share on a brisk October Saturday. I even hired a couple of bluegrass musicians to play live music for us. Some of my friends had a few handmade wares to sell, so I invited them to informally set up tables in the shade. I also have the amazing blessing of a couple of professional storytellers as friends. Gene and Peggy came along and regaled us with wonderful stories of the Old West, to round out the "cowboy campfire" atmosphere that had seemed to emerge. We set

up our spinning wheels and broke out our knitting needles and had a glorious day together. The day was a huge hit. And when something goes that well, my first instinct is to develop it further and make it even better next time. Which meant, there had to be a "next time."

We've repeated that party every year since. It has evolved into the "Share the Harvest" Fall Gift Market on the third Saturday in October, with almost two dozen vendors under large tents and small tents, our photographer friend Jennifer taking pictures of visitors with Shadrach the sheep, snacks, kids' crafts and more storytelling. Its growing success now requires the transformation of the north pasture into an attended parking lot. We've been able to attract a couple hundred shoppers to the farm each year to sample and support the work of some amazing local artists. These craftspeople are now part of the LRB's "extended family."

DFW Yarn Crawl—Connecting Yarn and Stitchers

Beyond the gates of the farm is the larger Dallas/Fort Worth (DFW) yarn community—knitters, crocheters, weavers and others who ply their crafts in little groups or in larger guilds or around yarn shop tables all over town. What would it be like to bring everyone together once a year for a citywide yarn event? I knew the yarn shop owners could use the financial support and that the yarn consumers would be happy to respond once they had a system to do so. Let's face it: none of us fiber artists needs too big of a push to organize a road trip with friends for yarn shopping!

I took a big, fat risk and began to organize the city's first Yarn Crawl.

Briefly, a yarn crawl is a coordinated yarn-shopping event in a given geographic area. "Crawlers" download a "passport" and get it stamped at each participating shop they visit over the duration of the crawl. These passports are surrendered at the end of the event, and winners are drawn from the entries, with prizes determined by how many shops a particular Crawler was able to visit. (Though some stitchers will play along with the crawl just for the pure sport of it, others need the lure of the prizes to fully commit to the game. We surely love free stuff.) Shops are encouraged to host their own special events during the crawl, luring shoppers in as they make their rounds. The goal is to motivate "yarnies" to visit stores outside of their usual routes, sample the amazing wares of the yarn shops, and have a great time

together. Many crawls feature a charity component—we will be collecting yarn and finished items for area NICUs and for a local women's prison knitting program.

I was actually shocked at the number of shops willing to give this idea a shot and back up their faith with cash. Some of the owners I had met, but many of them were off my beaten path, and I knew them only by name or reputation. To many of these ladies, I was basically a stranger. Thankfully, most of them recognized the business and publicity potential of a yarn crawl, and were willing to take the risk. I even had a bead shop contact me to participate; as it happens, beads and yarn go together rather well. The larger shops consented to give the project a try, and smaller shops were even more enthusiastic. The crawl leveled the playing field for them, and as word of the crawl got out, they often called me before I could call them. I was curious why no one else had organized a yarn crawl before in a metroplex the size of Dallas/Fort Worth. These events are going on all over the country in large cities, and have the potential to be the focal point of the yarncraft year for local knitters and crocheters. I realized that because I have a yarn-ish business, but am not really a yarn shop, I am a "safe" person to organize the event in an objective, non-biased way so that all the shops would get a fair shake. If any of the local shops had tried to organize it, the other shops might not have been willing to partner with a direct competitor.

Thirteen shops signed on for the experiment that first year, each one paying me to participate. The money would go toward building a website, printing buttons and my time to keep all the social media updated and to quarterback the thing. I enlisted a friend to help, and between us, we brainstormed the "rules." We made up a passport, printed buttons, and posters for the shops to advertise the crawl.

It's hard to be sure, but I wonder if a big plaster goat wasn't one of the biggest motivations for organizing that first yarn crawl? Earlier that year I had purchased a large mannequin goat from a guy on Craig's List. I mean, it was a huge *goat* with removable horns. How could I resist? But she was big and bulky and often in the way at the farm, with very little true purpose. I had been looking for a way to put this big novel piece to work for the farm, and a plan began to gel around "Geraldine," as we called her. (She was named after the goat in *The Goat in the Rug*, by Charles L. Blood, a book about a Navajo weaver that I read every year to our Farm Camp kids.) Several of our knitters and crocheters wanted to "yarn bomb" this goat and send her around to yarn shops to make "personal appearances." Geraldine had her own Facebook page, and served as sort of a mascot for the crawl that first year. She was decked, literally, horn-to-hoof with hand-stitched finery; she was quite the traffic stopper. We decided to auction her off at the big finale event at the farm. Fiber Circle, a yarn shop in Farmersville, Texas, won the auction, and Geraldine now occupies a place of honor in their front window. She brought a lot of whimsical attention to the crawl, and I easily got my investment back on her.

The first year of the crawl, we started slowly; we were making it up as we went. Getting sponsors turned out to be a bigger job than talking the shops into it, mostly because of the short lead time. Big companies prefer a little heads-up when making donations to festivals and events. They already get lots of requests for support, so they can be reticent to cough up cash. The first year, the yarn shops themselves donated prizes for the crawlers, and we made up baskets of goodies to give away. We had better success the second year with the large yarn and accessory companies, once we had a little experience under our belt. Participation jumped,

too, the second year. About 200 passports were returned and entered into the prize drawing at the end of the event. We know, statistically, that probably two or three times that number actually participated. Now, in the third year, we expect to do even better.

The guiding principle of the crawl for me as the organizer is to support the local yarn shops by encouraging yarn crafters to get out of their ruts and try out shops that they might not visit otherwise. We also wanted to encourage shoppers to give the stand-alone local yarn shops (LYSes) a fair shake when choosing yarn, rather than just running to the big box stores. We've scheduled the crawl at the beginning of the holiday crafting season, so that yarn consumers can stock up before they begin their gift knitting or crocheting. We've tried to make it very easy for stitchers to unload extra stash yarn that invariably builds up in their collections. First, giving to charity is just a wonderful thing to do; as we've discussed, yarnies love to give, and the charities will never run out of need. Second, "destashing" yarn helps us all feel a little less sated and more inclined to buy new yarn—which makes the shops happy.

In the second year, I wanted to draw attention to independent yarn artists—those yarn dyers, knit and crochet designers, and accessory makers who do not have their own brick and mortar shops. "Indies" could buy advertising space on a dedicated page of the yarn crawl website, and partner with physical shops for trunk shows or other feature events. As with the farm's Fall Gift Market, I am committed to promoting those business people who are working to make a living in the yarn arts, especially solo artisans without physical storefronts.

This process will continue to grow and mature over time, and my hope is that the DFW yarn community will continue to draw together in common support and celebration.

Farm Camp— Connecting Kids and Nature

For several weeks in June I get to put on a very special hat—my Teaching Kids hat. And I have to admit that I wasn't sure I'd be cut out for this at all. While I was trying to parent my young daughter, there were moms in my life who could run rings around me with the finger painting on the kitchen floor, and the cooking lessons with flour everywhere, and sidewalks covered with crazy chalk pictures. I called them "Play Dough Moms." They seemed to have infinite patience and a high tolerance for messes in the house. That, truth be told, wasn't me at all. Hopefully, my daughter has not been scarred for life.

In the following years I relaxed a lot. When I began thinking about Farm Camp, I realized I could design it to suit my own particular personality and tolerances. For instance, I've chosen to craft Farm Camp for kids between the ages of seven and ten because kids this age can take care of their own basic needs but have not yet developed that annoying eye roll. In fact, the kids we've had at Farm Camp so far have been, to a person, inquisitive, funny, adventurous, polite and eager to learn. I know it helps that we're studying alpacas and not algebra, but still, I appreciate their agreeable attitudes. It's a great experience for me, my adult helpers and the moms who get a few hours' peace during Farm Camp week.

I really enjoy working with the kids, but it does take quite a bit of energy, so we spend three hours together every morning, Monday through Friday. A comfortable balance

of indoor and outdoor activities make up the day, broken up by a brief snack time in the middle. Then at lunchtime I send them home lightly worn out. Me, not them. *I'm* lightly worn out. This schedule has been sustainable for me, the Introvert Who Prefers Adult Company If Given a Choice, and also perfect for the kids. I think they get everything they need from the experience.

The biggest challenge for me is to completely take off my Adult Shepherd hat and take myself back to a time when my fiber farm life—living among animals and really fun raw materials—was still new and novel. I ask myself to remember what things were interesting to me when I first started working with yarn and fiber. I also have to journey back to myself at the kids' same age, when I adored learning about nature and animals, especially when I could experience them up close, with my own hands. I have to reconnect with that "first love" wonder, so that I can guide the kids down the same wide-eyed path.

Some of the kids have returned to camp for several years in a row, so I try to freshen up the curriculum and list of activities, but some things would be missed if I replaced them. For instance, on Day One, I lead Shadrach, the Suffolk cross sheep, out into the yard and the kids take turns holding his lead rope and getting their pictures taken. We look at his teeth and feet and eyes, and feel his fleece. On Days Three and Four we work with wool: washing it, dyeing it in the sun, carding it, and felting it. The campers have even gotten a big kick out of helping me freshen up the chicken coops! Every day we work through a chapter book, usually about kids living on a farm.

I have had some very advanced kids come to Farm Camp (kids who can draw the water cycle from memory), but mostly, I can work my lessons around questions that come from adults who tour the farm from time to time. Kids and grownups both go all slack-jawed when I tell them that you don't need a rooster to get eggs and that egg-laying is more akin to ovulation than birth. They're amazed to learn that chickens have only one backside orifice out of which, and into which, everything passes. (The boys particularly love this factoid.) I have a million amazing livestock facts that get passed along to the parents on the trip home in the minivan, I'm sure.

I'm excited about next year's Farm Camp because, with the routine mostly settled for each day, I'm taking some time to sit quietly and think about cool things to add to the curriculum. I want to try the kids (both boys and girls) on loom knitting—let them make newborn hats for local NICUs. It's never too early to get kids helping other kids in the world. As usual, we'll grow some seeds in egg cartons, weave God's eyes, pet a chicken, lead an alpaca, groom a guard dog, and record everything in a personalized journal. Maybe the cicadas will emerge again as in past years, and we'll find newly morphed cicadas or the exoskeletons they've left behind.

Farm Camp succeeds if we entertain kids and develop in them a budding curiosity about how they get their food and textiles. I try to instill a basic respect for animals, personal property and each other. The adult volunteers come away having learned quite a bit and asking to come back again the following year. We barely touch the surface of amazing things to learn, and everyone knows we've left a lot of fun on the table. "Next year," we say. We'll try to get to those great lessons next year.

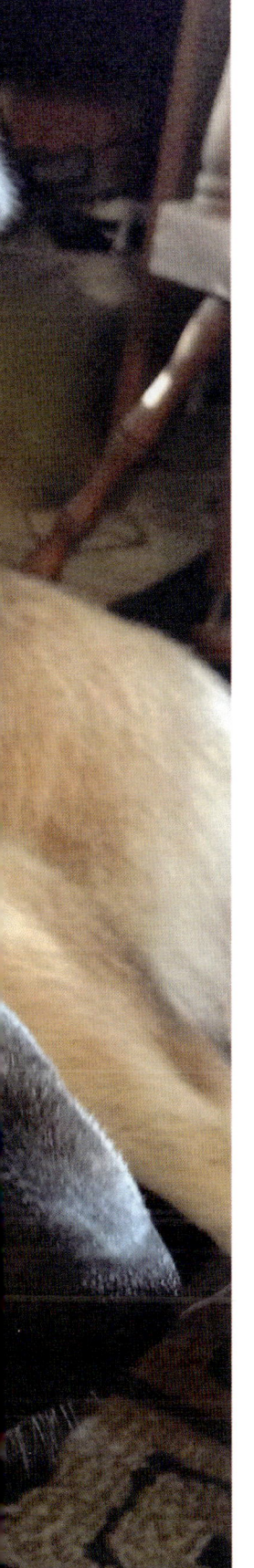

CHAPTER THIRTEEN

finding home

*I love the TARDIS-like quality of the barn when it seems like we can't fit one more person inside, and yet we do.
I love when there are just a few people left in the barn and conversations get deep and personal and the therapy goes beyond fiber therapy.
I love that I feel like I am part of a larger community at Jacob's Reward Farm.*

—RITA, BARNIE

Giving Away the Farm

A recently furloughed solder.

A kid who's lonely at camp.

A lover parted from her beloved.

A dove lost in a storm.

All these have one thing in common: they want to go home. I'm willing to bet that you've been so homesick at least once in your life that you thought you would die from it. I know I have felt that way. Home is not a geographic location. It's not a house. It's not even the place you came from, necessarily. It's the place where you know you belong. People know you there, and love you in spite of your flaws and weaknesses. That's where you're safe and secure and where your burdens are lifted.

It may seem a bit overblown to compare the atmosphere of a knitting group at a farm to the one place on earth where a person can feel loved and safe. But there are days when our sense of home cannot be found in the usual places. There are days when we feel anything but at home in our world, in our families, or even in our own skin. It's those times when we need each other. We need a place, even a silly, casual place where the niggling annoyances, the oppressive responsibilities, the unmet needs can be left outside for awhile, replaced with rest and joy and kindness.

This is what many of our friends have found at the Little Red Barn. It's what we hope to offer everyone who comes through the door.

Or who turns up at the gate…

Luther Comes Home

At first I thought he was a coyote—a tawny blur racing along the south fence of the property outside of the alpaca pasture, sending the guard dogs into apoplectic barking. But

Finding Home

then I realized he was too large and too yellow to be a coyote. Definitely a dog. That first day, I only saw him for a moment before he took off out of sight. When he returned a few days later he slowed his patrol of our borders to a lope and then a walk. I ran through the pasture waving my hands, trying to scare him off. He showed no interest in leaving. In fact, soon he was sitting and waiting for me to approach, wagging his tail. He may have been a stray, but he was not afraid of me, and not going anywhere. "Go home!" I yelled, which was not even a little bit effective. But it might have been a bit prophetic.

One evening as I was feeding the alpacas, he appeared right outside of the paddock, 15 feet away from me. I walked over to the fence, speaking to him quietly, and we locked eyes. The mouth was open and smiling, and the tail was wagging. He looked thin. I could tell he was hungry.

I pondered and went back and forth in my mind—I knew I was about to cross a line that I could never come back from. And then I did it. I poured him a small bowl of kibble and carried it down by the creek. I made him sit and wait before stepping back and allowing him to gobble up the food. And then I went about my other chores. After inhaling that food, he followed me around the property as I moved from north to south, feeding stock along the way. I didn't allow him into the sheep or alpaca pastures because I had no idea how he would behave. I was pretty sure the guard dogs would not react kindly to him. In fact, to this day, Judah growls and barks at Luther if he hangs out too close to the sheep fence. But this new dog took it all in stride. He waited for me outside the paddock gates, and when I finished one group of duties, he'd travel with me to the next. He quickly learned "get back" when I went through a gate, leaving him behind in the yard. He'd either wait patiently, panting in the grass, or he'd take

a little jaunt around the perimeter of the property, flushing out rabbits to chase and returning to my side when I came out of that pasture.

I really loved having a congenial companion to move around the farm with me, but I balked at the idea of adding another dog—another mouth to feed and to keep vetted—to the farm's collection. We already owned five dogs, a staggering number by my previous suburban standards. And here I was, considering Number Six. I chose a silly name for him, just as a reference, until I could actually decide if we should keep him or not. I called him Skippy.

Skippy had a bad limp. Everybody noticed it when he walked, though he could run like a bullet at the drop of a hat when inspired. If a car or truck sped by the property, Skippy bolted out the gate and down the road in hot pursuit. In fact, he even chased us if we left the property in one of our own vehicles. I wondered if, in addition to a heightened prey drive, he also might have some kind of separation anxiety. We tried all kinds of tactics to sneak out of the gate without setting off a dangerous game of chase. He would run on the left side of the vehicle, next to the driver's door, perilously close to the front tire. It didn't take a rocket scientist to figure out how he had gotten that limp. I also had a concern about his hips, which seemed narrow and sloped. His whole back end looked too small for the rest of his body. I chalked it up to his being undernourished, but I decided that if he came to stay, I'd have him X-rayed to see what was going on.

I was still really on the fence about whether or not to keep him, but he was boring his way into my heart with surgical precision. He seemed content to come for feeding and chore time, and then he'd disappear somewhere down the creek for the bulk of the day. Sure as clockwork, he'd be back for the next feeding, smiling and running alongside me as I moved through my duties. Occasionally he'd be "late" for chores and I caught myself worrying that he'd been hit by a car—or worse yet, claimed by somebody else. I missed him when he was gone. Then, as soon as I had taken a deep breath and let go of the whole idea of keeping him, he'd come loping up to lick my hand and stick to my hip as I worked.

Soon, he was hanging around longer and longer during the day, sacking out on the porch or in the shade between rounds of chores. At night, we could hear him barking in the yard, protectively circling the house and sounding ferocious. That's a wonderful job for him, I thought. We might keep the coyotes at bay once and for all with Skippy on the job—a job he had self-selected, and for which he had needed no training. This was great. Maybe … we should keep him. He certainly needed a place to belong.

Everything at the farm is public news on our Facebook page, and Skippy was attracting quite a fan club. Even through my iPhone photos, folks immediately picked up on his winsome personality and could see the practicality of keeping him around. I was getting significant pressure to make him part of the farm's permanent family. Any arguments for handing him to a shelter began to evaporate. He seemed to be made for the farm, and the farm for him.

If Skippy was to truly join the farm, one important trait he had to have was a tolerance for visitors, especially very young ones. We have people coming and going all the time and I needed to trust that this loose dog was not going to be a biting or jumping risk. One day, a group of homeschooling moms and their toddlers visited the farm, keen on enjoying our animals up close. I had the dog on a leash, just to have some control in case a kid got too grabby or invasive, forcing him to react defensively. But Skippy took it all in stride. He acted like he had nothing better to do than to have several

little kids kiss him and hug him and stroke his body in that choppy, ham-fisted way that they can. I watched him like a hawk for any sign of discomfort or fear, but he stayed perfectly calm and relaxed—not just tolerating the kids, but soaking up every tidbit of affection. It was a mutual love fest between dog and kids. The moms snapped photos and thanked me for letting the kids enjoy my wonderful dog. I just smiled, nodded and let out a huge sigh of relief.

That did it. The dog was here to stay.

I named him Luther. Maybe it was the combination of his big protective bark, his vaguely German Shepherd-ish looks, and my admiration for the sixteenth-century German theologian Martin Luther, but that's what I came up with, and it stuck pretty quickly. He was official. He had won the hearts of all the Barnies and Facebook friends, and if I had so much as *thought* about handing him over to the pound, I would have been in big trouble with the community.

Almost immediately, our free dog began costing us money. I was working on the computer one morning and heard a yelp of pain from outside. I ran out onto the porch to see a city truck idling on the road in front of the house and Luther limping toward me with his right hind leg not touching the ground. The driver of the truck apologized profusely, very concerned about hurting the dog. "He ran right out of nowhere and I couldn't stop!" I assured him that Luther had a problem with chasing, and that it wasn't his fault. After I promised the driver that everything was OK, he apologized again and drove away. Then I noticed that Luther's foot was bleeding, and had begun to swell pretty badly. Dang. This was going to require a trip to the vet. My friend Michelle was scheduled to come knit in the barn with me that morning, but instead, found herself taxiing Luther and me to the animal hospital. There we found out that his leg was not broken but he had some swelling and abrasions. We scanned him but found no microchip. The doctor recommended letting this injury heal for a few weeks before we brought him back for the full spa treatment: vaccinations, chipping, neutering and hip x-rays. Luther had landed in tall cotton.

While he healed, Luther made the acquaintance of the LRB gang, and it was mutual love at first sight. Luther loved nothing more than to stretch out on the floor of the barn, surrounded by knitting, crocheting, spinning, snacking and chatting. The ladies oohed and aahed and made over him like he was a movie star. He supervised every entrance and exit, tasted most of the snacks and soaked up every pat and bit of praise. Luther was home.

I crossed the final line of resistance and allowed him into the house when he came home from his Sports Package Removal surgery. I just wanted to be sure his healing went according to plan. His transition to an indoor/outdoor dog went so smoothly I wondered if he had indoor experience from wherever it was that he came from. He made friends with the cranky old indoor Corgi, and eventually graduated to irritated tolerance from Clare the cat. He never had accidents in the house, and simply used his huge bark to let us know he wanted out.

It was as if he had been destined to call Jacob's Reward Farm home.

I really think the Barnies love Luther so much because they unconsciously see themselves in him. I know I do. We were all looking for a place to belong, to fit in, to be accepted and loved and to use our gifts and talents for the good of the community. The Barnies have championed Luther's cause more zealously than any animal that has ever wound up here before, because he is one of us. Luther is a Barnie with fur. He is home.

CHAPTER FOURTEEN

Community Killers

I've always been with "people like me" but sitting in the LRB on my first visit was a Wiccan, a Muslim, a Christian and an atheist, and we all got along, and we all had so much in common, and we all genuinely cared about how the others were doing and our differences made the conversation richer. It was spectacular.

—KATE, BARNIE

Giving Away the Farm

Books and books have been written on the complicated dynamics of group interaction. That's way above my pay grade. I'd like to talk to you, Yarn Crafter, about what to seek and what to avoid when you go looking for a place to rest, create, and develop friendships with others who love what you love. The different attributes you'll encounter are polar opposites, and pretty easy to recognize if you watch closely.

For instance, if we need acceptance, we must disallow judgment. If we need safety, we must avoid criticism. If we need inspiration, we must prevent negativity. If we need wonder and awe, we must work against cynicism. If we're striving for truth, we must reject deception and insist on integrity.

A cohesive group is very often a more creative group. And creative is what we're all about, right? So, what glues us together? A shared love of fiber and yarn, a shared journey of learning skills, a shared sense of accomplishment in creating stuff out of string (even if we just follow the pattern), shared challenges of creating in the midst of everyday pressures and distractions, shared concerns about measuring up and a shared kinship as human beings in search of a tribe.

Art begets art, and a strong artist can be challenged and encouraged by other artists. But first, you have to find (or create) a place where artists can flourish. Artists are, almost by definition, people living out on the edges of culture—stretching boundaries, testing limits, rejecting dogma. As a result, we often feel alone and vulnerable. Artists need acceptance, freedom, safety, fun, humor, challenge and room to experiment, or we get worn out and find our passions dulled.

Think about it. Are you more likely to come up with a wonderful idea in a group of folks you trust and who appreciate your work, or in a group of strangers whose motives you can't discern? When you hear people next to you gossiping about someone else, does that make you more or less likely to share your exciting brainstorm?

If you see someone try to come into the group who is met with silence, rolled eyes, or just vague judgment, what does that tell you about the group culture? Do you feel more or less willing to express radical ideas, or dress unusually, or not be at your best? Are you accepted for who you are, no matter what?

The Role of the Leader

Do groups even need a leader? At the LRB, I became the default leader because it was my studio at my farm. I started out as the host, and naturally stepped up to be the hub for communication and planning activities. It's in my personality to guide and set the tone on my own territory. But not all groups have or need lone leaders. This responsibility can be shared by several people who hold similar goals and strategies for the group. Some folks might prefer the title "facilitator" to "leader," and that works, too. But most groups find that just for practical reasons, they need someone to act with some kind of leadership. Somebody has to make decisions.

I hesitate to set out guidelines for leaders because I know I'm painting a tough set of ideals for myself, but that is the hazard of the job. The bar is high, because the stakes are high. Fortunately, one of the great things about a healthy group is that grace abounds, and we always get forgiveness and do-overs when we inevitably mess up.

The leader's goal is to provide an environment where art can flourish. Anything that puts a damper on these things needs to be rooted out and removed. But leaders must be kind and gentle even in the ways they deal with negative

influences. To maintain a positive, nurturing atmosphere, the leader must sometimes deal with problems quietly and privately, away from the group. Humiliation and shaming have no place in the Culture of Kindness.

Avoid groups where the leader is more interested in control than creativity, or who insists on always being the center of attention, or who cannot compliment or praise others or who wields power selfishly. You will always be just a cog in this leader's wheel. You will not be valued as anything other than a means to her end. She needs an audience, and you are it. Walk away.

Healthy leaders, on the other hand, serve their group rather than insisting that they are served. They share responsibility, lavish praise, and encourage vulnerability by being vulnerable themselves. They place the needs of their group above their own. They are humble enough to ask for and graciously receive help. Leadership is hard not because one must be loud and strong, but because one must be a humble servant. A good leader shows kindness and grace, knowing she herself will need those things from the group from time to time.

Challenging Members

I'm pleased to say that in our particular case, I've hardly ever had to go to any dramatic lengths to deal with someone who came into the group and started causing trouble. The whole group is on board with our Culture of Kindness, and I believe that we actually make the environment unfriendly to unfriendliness. Like good bacteria in the body, we surround any hint of nastiness and overpower it with grace.

I was involved with a different group who once had to deal with a challenging personality. She became so offensive and

disruptive, and oblivious to both subtle and overt attempts at correction, that the group actually had to temporarily disband and regroup secretly in a new location to rid themselves of her. It was the kindest thing they could have done. They weren't responsible for her remediation or reformation—they were a knitting group not a therapy group after all. But they did what they had to do in order to keep the group safe. Once the

group got back together, they could resume their function as a place of personal safety and artistic encouragement.

To be clear, the Culture of Kindness does not stamp out all fun and raucousness. Not in the least. The LRB can rock with crazy, rowdy laughter and sharp wit. But there's nothing mean about it. There's nothing invasive or threatening. Our conversation is full of texture and nuance, hilarity and grittiness. We can enjoy all these things without resorting to sniping or bad manners.

The LRB community is not huge. We can only fit 12–15 people in the barn at any one time. Like many fiber groups, we sit in a circle with our needles, hooks and wheels, and can look each other in the eye. Depending on specifically how many people we shoehorn in, we may be sitting elbow to elbow. We do not have the veil of anonymity that might allow for subtle negative vibes to grow into sarcasm or division. Unlike the anonymity of road rage on the freeway where your fellow travelers are just faceless annoyances to be yelled at or flipped off, the people around the LRB are clearly humans with feelings. We have the benefit of hearing voice inflections and reading body language in our conversations. We can communicate heart to heart, and we are called to a more stringent code of behavior. We have accepted the challenge to live as friends in close quarters. And the bonds have grown so that they hold even outside the walls of the Little Red Barn.

My wish for you, Yarn Ninja, is that you'll find a community that fits you like a glove—a place where you are welcomed, encouraged, valued, and challenged to be the best yarn or fiber artist you wish to be. I pray that this community will support you not only in your craft, but in your personal life, as you learn to both give and receive in the company of trusted friends.

Community Killers

If you have a community like this, celebrate it and help it grow strong. You are rich with a rare and precious treasure.

If you do not have a community like this, start one. Gather knitters, crocheters, spinners and weavers around you. Commit to caring for them and creating a nurturing environment for them. Serve them and love them. The joy that comes back to you will fill your heart to the brim.

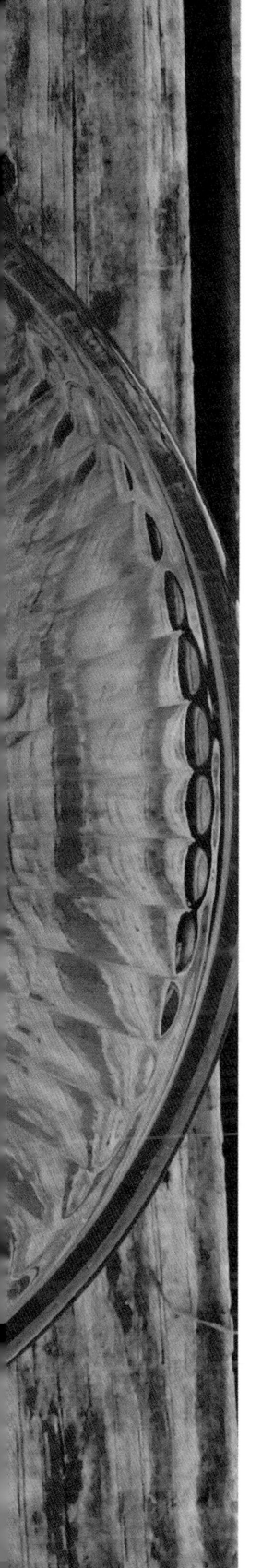

CHAPTER FIFTEEN

My Parting Promise

When Cindy and I met, I knew we were in the same tribe. Our shared loves drew me toward her and eventually the land she and Ted have acquired. On the day I first came to the farm packing deep needs and a golden pear which we shared for lunch, I pulled weeds, smelled the earth, connected with sheep and a blossoming farm that said "Come!"

—LAURIE M., BARNIE

Giving Away the Farm

As the Steward of this farm, my goal is to make it more available and accessible to the people who need it: artists, consumers, families, aspiring fiber farmers, preachers, land-locked city dwellers, and Truth seekers. If you can think of a way that this special place can fill a need, I'd like to hear about it. Please contact me through our website—www.jacobsreward.com.

And because the need for this space, this safety, this refuge, comes more often than just on the appointed days, I want you to know, that if you find yourself in need of some quiet Barn Time, just email me.

I'll leave the light on for you.

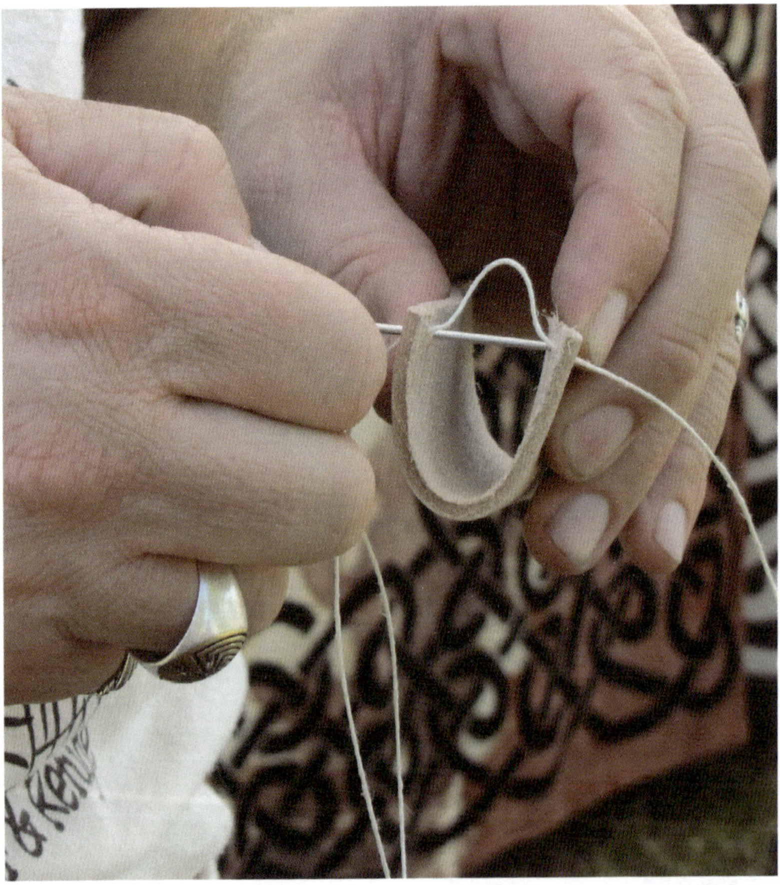

Endnotes

1 John 21:15-17.
2 Matthew 25:14-30.
3 Genesis 30.
4 http://www.goldingfibertools.com/ringspindles/custom.
5 *One Colour, 54 Fiber Artists Interpret the Theme*, Cyra Lewis, 2014.

About the Author

Photo by Jennifer Jurek Photography

Cindy Telisak is a lifelong communicator, telling important stories in her serial careers as an actor, teacher, radio personality, photographer, copywriter, blogger, lyricist and now in her new role as an author. She finds bountiful inspiration on this farm she shares with her husband and daughter, and aspires to help other women entrepreneurs tell their unique stories through compelling branding and marketing.

Made in the USA
Lexington, KY
05 December 2014